Canadian Political Science Association

CONFERENCE ON STATISTICS 1961

T0326656

Canadian
Political Science
Association

CONFERENCE
ON STATISTICS
1961

Sir George Williams University

Montreal, Canada

PAPERS

Edited by

WM. C. HOOD and JOHN A. SAWYER

UNIVERSITY OF TORONTO PRESS

© *University of Toronto Press 1963*

Reprinted 2017

ISBN 978-1-4875-9179-3 (paper)

Programme

CANADIAN POLITICAL ASSOCIATION

CONFERENCE ON STATISTICS

Sir George Williams University, Montreal

June 11 and 12, 1961

June 11

CHAIRMAN: Gideon Rosenbluth, Queen's University (now at University of British Columbia)

"Inter-Industry Estimates of Canadian Imports, 1949-1958"
T. I. Matuszewski, University of British Columbia (now at University of Montreal)
Paul R. Pitts, Dominion Bureau of Statistics, Ottawa
John A. Sawyer, University of Toronto

"A Regional Table of Inter-Industry Flows, Province of Quebec, 1949"
André Raynauld, University of Montreal

DISCUSSANTS: S. J. May, Department of Trade and Commerce, Ottawa
T. M. Brown, Royal Military College of Canada (now at Queen's University)

CHAIRMAN: A. D. Scott, University of British Columbia

"The Standard Industrial Classification and the Standard Commodity Classification"
Neil L. McKellar, Dominion Bureau of Statistics, Ottawa

"Forthcoming Changes in the Census of Industry"
V. R. Berlinguette, Dominion Bureau of Statistics, Ottawa

DISCUSSANTS: M. C. Urquhart, Queen's University
D. W. Slater, Queen's University

CHAIRMAN: D. N. Solomon, McGill University

"La Détermination des zones agricoles sous-marginales"
Gérald Fortin, Laval University

"Regional Aspects of Canadian Labour Mobility"
H. F. Greenway and G. W. Wheatley, Dominion Bureau of Statistics, Ottawa

DISCUSSANTS: J. Henripin, University of Montreal, and N. Keyfitz, University of Toronto

Sylvia Ostry, McGill University (now at University of Montreal)

June 12

CHAIRMAN: J. J. Deutsch, Queen's University

"The Characteristics of Persons Looking for Work: A Survey of Registrants with the National Employment Service"
R. W. James, Department of National Defence, Ottawa

"Trends and Fluctuations in the Post-War Canadian Labour Market"
F. T. Denton, Dominion Bureau of Statistics, Ottawa

DISCUSSANTS: W. Donald Wood, Queen's University
D. G. Hartle, University of Toronto
A. Asimakopulos, McGill University

CHAIRMAN: Yves Martin, Laval University

"Flow of Migration among Provinces in Canada, 1951-1961"
Yoshiko Kasahara, Dominion Bureau of Statistics, Ottawa

"Population Migration in the Atlantic Provinces"
Kari Levitt, University of Toronto (now at McGill University)

DISCUSSANTS: K. A. H. Buckley, University of Saskatchewan
Z. W. Sametz, Department of Citizenship and Immigration, Ottawa

Preface

THIS VOLUME contains some of the papers and comments offered at the Second Conference on Statistics sponsored by the Canadian Political Science Association and held at Sir George Williams University, Montreal, on June 11 and 12, 1961.

The programme committee that arranged the conference consisted of D. W. Slater (chairman), A. E. Safarian (vice-chairman), Nathan Keyfitz, James Hodgson, and Yves Martin. The full programme of the Conference is printed above.

Not all of the papers given at the Conference are published in this volume. We have included, however, all of the papers which have not already been published, or published in substantially similar form, in readily available publications. We list herewith the publications in which the papers not included in this volume have appeared:

André Raynauld, "A Regional Table of Inter-Industry Flows, Province of Quebec, 1949" is covered in chapters 3 and 4 of his book entitled *Croissance et structure économiques de la Province de Québec* and published by le Ministère de l'Industrie et du Commerce, Province de Québec, 1961.

V. R. Berlinguette, "Forthcoming Changes in the Census of Industry" is published in the *Canadian Statistical Review*, July, 1961, pp. i-vii, and reprinted in Dominion Bureau of Statistics publication no. 13-518, *Selected Articles on DBS Statistical Activities* (1962).

N. L. McKellar, "The Standard Industrial Classification and the Standard Commodity Classification" is published in the *Canadian Statistical Review*, May and June, 1961, pp. i-viii and viii-xii respectively, and reprinted in Dominion Bureau of Statistics publication no. 13-518, *Selected Articles on DBS Statistical Activities* (1962).

R. W. James, "The Characteristics of Persons Looking for Work" published in the *Proceedings of the Special Committee of the Senate on Manpower and Employment*, no. 4, Jan., 1961, and no. 8, appendix, Feb., 1961 (Ottawa, Queen's Printer).

We have not included comments on papers not published in this volume. The comment by D. G. Hartle on the paper by R. W. James appeared in the *Canadian Journal of Economics and Political Science*, vol. XXVIII, May, 1962, pp. 254-62.

We have included comments that were received by the editors on papers published in this volume.

On behalf of the Statistics Committee of the Canadian Political Science Association, we wish to acknowledge gratefully the assistance of the Social Science Research Council of Canada which made this publication possible.

WM. C. HOOD and J. A. SAWYER

University of Toronto
May, 1962

Contents

Regional Aspects of Labour Mobility in Canada, 1956-1959

H. F. GREENWAY and G. W. WHEATLEY

THIS PAPER is concerned with job changes within the Canadian population covered by unemployment insurance. Currently [1961] the insured population numbers more than 4,000,000 persons. It represents about two-thirds of the Canadian labour force, and the insured employed are approximately four-fifths of all paid workers. At any point in time the insured population includes both employed and unemployed persons, and between points in time people are constantly entering and leaving. Those who leave the insured population may be withdrawing from the labour force or changing to some form of uninsured employment, and conversely those who enter may come from outside the labour force or from uninsured employment.

At annual intervals, records of persons in the insured population can be used to provide a census of these persons. Although not primarily designed for labour mobility studies, such records are quite useful for this purpose, since they can be used to trace year-to-year changes in location, industry, and occupation of insured persons. The principal limitation of the information is that it relates to personal circumstances at only one point in time each year—when unemployment insurance books are renewed in May or early June. If a person happens to be unemployed at book renewal time, this fact is recorded along with his location and personal characteristics. It is therefore possible to distinguish between the unemployed and the employed, and to examine movements and personal characteristics of both groups. For the employed, it is also possible to extend the study of mobility to include job changes involving different industries and occupations. The distinction between the employed and unemployed, of course, is not clear-cut. Summer is a time of low unemployment, and undoubtedly many persons employed in May and June will be omit unemployed later in the year, omit particularly between December and March.

Two geographic units of mobility analysis have been used: the province, and the local office area of the National Employment Service of the Unemployment Insurance Commission. There are 208 of these local office areas, and they have populations differing widely in size and in urban-rural composition.

In addition to local office area, industry, and occupation, book renewal records indicate the age, sex, and marital status of insured persons. They also identify individuals by insurance numbers which are assigned for life, and this provides the means of relating records for the same person over a period of years. The num-

C.P.S.A. Conference on Statistics, 1961, *Papers*. Printed in the Netherlands.

bers are also useful in drawing samples, and for this study a one per cent sample of approximately 40,000 records each year was secured by drawing every number ending in 34. Each year newly issued numbers ending in 34 were added to the sample, thereby keeping it representative of the insured population.

Different methods of calculating mobility rates from these data have been used for special purposes, but the main part of the study is concerned with rates calculated after matching records for pairs of years, and then expressing job changes as percentages of the possible job change which could have been recorded. Some idea of the nature of the records which were used can be obtained from a simple diagram (Table 1) in which five symbols are used. OS stands for original sample records, NE represents a record for a new entrant to the insured population, NJ indicates a change to a new job, SJ indicates no change of job, and NR indicates that there was no record for the year in question. This might occur when a person withdrew from the labour force or took employment that was not insurable. Also, in a small number of cases it might result from non-response on the part of employers or from loss in handling of the records.

TABLE 1

Sample Records

	1956	1957	1958	1959
Case 1	OS	SJ	NJ	NJ
2	OS	NR	NJ	SJ
3	OS	NJ	SJ	SJ
4		NE	NJ	NJ
5			NE	SJ
6				NE
Total possible job changes		2	3	5
Job changes made		1	2	2

Because there was no record for Case 2 in 1957, the 1956-57 comparison was limited to two cases. In 1958 this person was back in the sample in a new job, but his record could not be used in the 1957-58 comparison. Thus the 1957-58 comparison was limited to cases 1, 3, and 4. This type of counting, in effect, assumes that persons for which there was no record in a given year were not in the insured population. The proportion of persons who withdrew temporarily from the insured population was about 12 per cent in the years 1956 to 1959.

It has been noted that records which indicate an individual's work status at one point in time each year can record only one job change a year, so that intermediate changes or periods of unemployment will not be recorded. Likewise there will be no record of a change which involves a new employer without any change in local office area, industry, or occupation. These are factors which are likely to reduce annual rates of changes below their actual levels.

At the present time the insured population excludes some employees such as hired farm help who may be presumed to have greater than average mobility, and others such as government employees who have less than average mobility.

In the following sections the term *total mobility* refers to job changes involving one or more of the three variables, local office area, industry, and occupation. A simple change involving only one variable, such as local office area, rates equally with one in which all three recorded variables were involved. The terms *interprovincial mobility* and local office area mobility are also used. These apply to any job change involving a change of province or of local office area. Total mobility rates, of course, are much higher than rates which involve only geographical change, and local area rates which reflect many short moves are considerably higher than provincial rates. The differences between *provincial* mobility rates and *local office area* rates for the same province record *intraprovincial* rates of movement. *Industrial* and *occupational* mobility rates reflect job changes recorded by three-digit codes which provide rather fine classification groups of industries and occupations.

SUMMARY OF FINDINGS

1. Total mobility rates for the population covered by unemployment insurance in Canada declined between 1956 and 1959. The decline between 1957-58 and 1958-59 was unusually pronounced, and probably reduced mobility rates to the lowest levels in the post-war period. The experience of the 1956-59 period suggests that mobility rates react fairly quickly to a substantial increase in unemployment. However, they failed to respond to the minor improvement in employment conditions during 1958-59. On the contrary, all types of mobility rates, regional, industrial, and occupational, fell sharply during that period, when a small gain in employment failed to reduce high unemployment levels by an appreciable amount.

2. The Maritime provinces and the three most westerly provinces were areas of higher than average labour mobility, while mobility in Newfoundland, Quebec, Ontario, and Manitoba was lower than average.

3. Differences occurred between interprovincial and intraprovincial mobility in 1957-58 when unemployment increased sharply. Interprovincial movement increased while intraprovincial movement declined. However, both long and short moves were less frequent in the following year.

4. Provincial rates of outflow and inflow changed substantially from year to year, and in most cases both flows contracted appreciably in 1958-59.

5. There was evident that unemployment is a major factor contributing to labour mobility. Also there appeared to be differences in the character of voluntary and involuntary mobility. All mobility rates for the unemployed were higher than for the employed. However, occupational mobility was comparatively high for the employed, while industry and local area mobility were relatively more important among the unemployed.

EMPLOYMENT AND UNEMPLOYMENT
DURING THE SURVEY PERIOD
(May 1956–May 1959)

Since it is generally recognized that employment opportunities exert a strong influence upon labour mobility, a very brief outline of employment and unemployment changes during the survey period will provide relevant background material against which to examine changes in mobility. Several facts about this period stand out very clearly.

In 1957-58, the central year of the three-year period, employment failed to increase. This was only the second year in the post-war period in which any decline in employment had occurred, and unemployment in the same year increased by 180,000.

Although employment increased in the following year, 1958-59, by 100,000, this was a comparatively small gain, and it was accompanied by a very small drop of 34,000 in the number of unemployed. Thus, in May, 1959, the number of unemployed, at 355,000, was only a little lower than in May, 1958, and it remained twice as high as in May, 1956. Comparable levels of unemployment had not been experienced since the winter of 1939-40 (see Table 2).

The years 1956 to 1959 witnessed a transition from a period of comparatively low unemployment to one of high unemployment. It is not surprising, therefore, to find regional differences in mobility behaviour during the middle of this period, and more consistent behaviour from region to region during the final year.

TABLE 2

Employment and Unemployment in Canada, May, 1956, to May, 1959

Area	May 1956	May 1957	May 1958	May 1959	May-to-May changes		
					1956-57	1957-58	1958-59
	all figures in thousands						
	Employed						
Atlantic	487	499	483	491	+ 12	— 16	+ 8
Quebec	1,523	1,573	1,575	1,589	+ 50	+ 2	+ 14
Ontario	2,079	2,168	2,144	2,177	+ 89	— 24	+ 33
Prairies	988	1,002	1,025	1,043	+ 14	+ 23	+ 18
British Columbia	486	519	504	531	+ 33	— 15	+ 27
Canada	5,563	5,761	5,731	5,831	+198	— 30	+100
	Unemployed						
Atlantic	33	39	64	66	+ 6	+ 25	+ 2
Quebec	78	76	139	143	— 2	+ 63	+ 4
Ontario	37	59	109	87	+ 22	+ 50	— 22
Prairies	14	16	34	29	+ 2	+ 18	— 5
British Columbia	13	19	43	30	+ 6	+ 24	— 13
Canada	175	209	389	355	+ 34	+180	— 34

LABOUR MOBILITY AMONG THE EMPLOYED, 1956-1959

Annual mobility rates for the employed were based upon pairs of records for persons in insured employment at consecutive unemployment insurance book renewal periods. Thus, 1956-57 rates represent the number of recorded job changes between May, 1956, and May, 1957, per 100 persons in insured employment during both of these periods.

The first part of this section utilizes data based on samples from all of Canada of persons employed at book renewal time in order to give some indication of the complexity of labour mobility, and of the relative proportions of job changes between local office areas, industries, and occupations. This is followed by an examination of provincial mobility rates, and of interprovincial movement among insured workers. Finally there is a brief note on intra-provincial movement.

1. Canada

A sharp decline in labour mobility rates occurred between 1956 and 1959, with the greater part being concentrated in 1958-59. Distributions of job changes according to the various possible combinations of the three mobility variables, local office area, industry, and occupation, showed that changes in occupation and industry, were much more frequent than movements to different local office areas. In 1958-59, nearly 50 per cent of changes were to different occupations in the same industry and area (see Table 3).

TABLE 3

Distribution of Job Changes according to Mobility
Variables Involved

Year	Total mobility rates	Percentages of job changes involving						
		Local office area only	In-dustry only	Occu-pation only	Local office and in-dustry	Local office and occu-pation	In-dustry and occu-pation	Local office in-dustry and occu-pation
1956-57	54	6.1	17.7	39.5	3.4	2.9	22.0	8.4
1957-58	50	7.3	15.1	44.1	3.6	3.5	19.2	7.2
1958-59	41	4.2	16.8	48.6	3.4	2.2	18.1	6.7

As mobility declined, a slightly larger proportion of changes involved only one of the three variables. Proportions of changes involving local office area and industry declined relative to those for occupation. About three-quarters of job changes between 1956 and 1959 involved shifts in occupation. For industry and local office area the corresponding fractions were close to one-half and one-fifth (see Table 4).

TABLE 4

Relative Importance of Individual Mobility Variables

Year	Percentages of job changes involving			
	Only one variable	Local office area	Industry	Occupation
1956-57	63	21	51	73
1957-58	66	22	45	74
1958-59	69	16	45	76

Mobility rates for industry declined fairly evenly between 1956-57 and 1958-59. Local office area rates for Canada showed no change between 1956-57 and 1957-

58, but fell sharply in the following year. Differences in mobility of the insured population at progressive stages of high unemployment will be observed in all of the data subsequently presented. The general pattern followed that of total mobility, with a moderate decline in the first year followed by a larger one in the second (see Table 5).

TABLE 5

Canada Mobility Rates for the Employed, 1956-59

Year	Total	Local office area	Industry	Occupation
1956-57	54	11	28	39
1957-58	50	11	23	37
1958-59	41	7	18	31

2. Provinces

a) Total Mobility Rates. Apparently provincial differences in industrial structure have less to do with differences in labour mobility than other factors, such as geography and economic conditions. Industry mobility rates for the provinces behaved in a quite uniform manner. Seemingly more sensitive to rising unemployment levels than other types of rates, industry rates declined, almost without exception, by substantial amounts from 1956-57 to 1957-58, and again the following year.

There was much less consistency in the behaviour of local office area rates. Some of these increased slightly between 1956-57 and 1957-58, while others showed no change, and a few showed small declines. However, in every case they fell appreciably between 1957-58 and 1958-59. It should be noted that these rates reflect both intra- and inter-provincial movement. It will be shown later that intraprovincial rates like industry rates declined in 1957-58, while interprovincial rates generally remained firm or increased.

Occupational mobility rates for most provinces showed small declines in 1957-58, and proportionately smaller declines than those for local office area and industry rates in 1958-59.

In examining rates for individual provinces, the influence of geography and economic conditions appeared intermingled to a considerable extent. Newfoundland rates were low, yet those for Prince Edward Island and British Columbia were high. Manitoba rates, which were low but sensitive to economic change, differed appreciably from those of its closest neighbours. Saskatchewan rates were high and sensitive, while Ontario rates were comparatively insensitive. Quebec rates were close to those for Ontario, but rather more sensitive. Local office area rates for New Brunswick also were quite sensitive, being relatively high in 1956-57, but about average for 1958-59. Industry rates for Nova Scotia were slightly less sensitive than those for New Brunswick, although occupational rates for both provinces were about the same. Most Alberta and British Columbia rates moved at comparable levels which were generally above those for Canada (see Table 6).

b) Provincial Labour Mobility Rates Converted to Ratios of Canada Rates. Industry and occupation mobility rates were much higher than local office area rates, and dispersion in provincial rates for industry and occupation was fairly substantial. Thus, one might easily be misled as to the relative importance of provincial differences in the three types of rates. To place these differences in proper

TABLE 6

Provincial Labour Mobility Rates, 1956-1959

Province	Total			Local office area			Industry			Occupation		
	1956-57	1957-58	1958-59	1956-57	1957-58	1958-59	1956-57	1957-58	1958-59	1956-57	1957-58	1958-59
Nfld.	50	50	39	8	5	3	26	22	16	35	41	30
P.E.I.	61	53	40	16	17	*	33	27	19	36	29	27
N.S.	54	54	37	14	17	6	29	23	17	38	38	27
N.B.	59	49	34	16	13	7	33	26	16	39	33	27
Que.	56	50	43	10	9	5	28	21	18	42	38	34
Ont.	51	47	40	10	10	7	25	21	18	38	35	30
Man.	54	51	31	10	12	5	29	23	16	35	36	22
Sask.	59	63	39	19	20	9	31	30	17	41	43	30
Alta	58	55	47	13	11	13	35	30	21	40	39	33
B.C.	57	54	44	14	14	7	32	24	20	40	39	33
Canada	54	50	41	11	11	7	28	23	18	39	37	31

* No estimate.

perspective, ratios were calculated by dividing all provincial rates by the corresponding Canada rates. Besides placing relatively high industry and occupation and the lower local office rates upon a common footing, ratios also removed any arbitrary differences in rate levels introduced by different classification systems for the three variables. As an example, there are only 208 local office areas, whereas three-digit classifications of industries and occupations produce substantially larger numbers of groups. However, ratios may be deceptive when comparing data for a series of years. As an example, the local office area ratio for Alberta in 1959-59 showed a sharp rise over the two preceding years, a rise accounted for mainly by a sharp drop in the Canada rate for 1958-59.

Provincial mobility rates expressed as ratios of Canada rates demonstrate that interprovincial differences in local office area rates are much greater than those for occupation and industry. They also show very clearly the substantial year-to-year differences in rate changes which occurred in different provinces.

Table 7 shows up very clearly the relatively high mobility of the Maritime provinces and western Canada and the lower mobility of Newfoundland, Quebec, and Ontario. It also points to several unusual occurrences which do not appear to be due to sampling error. One of these, the well-maintained rate of local office area mobility in Alberta during 1958-59, has already been mentioned. It will be shown in a later table that rates of outflow from Alberta actually increased gradually from 1956-57 to 1958-59. The relatively sharp contraction in all three types of mobility rates for Manitoba in 1958-59 is another unusual feature of interest.

c) Interprovincial Movement of Insured Workers, 1956-59. Patterns of interprovincial movement were established for each of the three yearly periods commencing May 1956, 1957, and 1958. Both inflow and outflow rates were calculated, with the latter being taken to represent interprovincial mobility rates. Flow patterns for all provinces show consistency from year to year, although the samples were small, particularly for the Atlantic provinces and Saskatchewan. Wage-earners

TABLE 7

Provincial Mobility Ratios, 1956-1959
(Canada rates equal 1.00 in all cases)

Province	Total			Local office area			Industry			Occupation		
	1956-57	1957-58	1958-59	1956-57	1957-58	1958-59	1956-57	1957-58	1958-59	1956-57	1957-58	1958-59
Nfld.	.9	1.0	1.0	.7	.5	.4	.9	1.0	.9	.9	1.1	1.0
P.E.I.	1.1	1.1	1.0	1.5	1.6	*	1.2	1.2	1.1	.9	.8	.9
N.S.	1.0	1.1	.9	1.3	1.6	.9	1.0	1.0	.9	1.0	1.0	.9
N.B.	1.1	1.0	.8	1.5	1.2	1.0	1.2	1.1	.9	1.0	.9	.9
Que.	1.0	1.0	1.1	.9	.8	.7	1.0	.9	1.0	1.1	1.0	1.1
Ont.	.9	.9	1.0	.9	.9	1.0	.9	.9	1.0	1.0	1.0	1.0
Man.	1.0	1.0	.8	.9	1.1	.7	1.0	1.0	.9	.9	1.0	.7
Sask.	1.1	1.3	1.0	1.7	1.8	1.3	1.1	1.3	.9	1.1	1.2	1.0
Alta	1.1	1.1	1.2	1.2	1.0	1.9	1.3	1.3	1.2	1.0	1.1	1.1
B.C.	1.1	1.1	1.1	1.3	1.3	1.0	1.1	1.0	1.1	1.0	1.1	1.1
Canada	1.0	1.0	1.0	1.0	1.0	1.0	1.0	1.0	1.0	1.0	1.0	1.0

* No estimate.

form a much smaller proportion of the labour force in Saskatchewan than in other western provinces, and the insured population of Saskatchewan is less than half the size of corresponding population in neighbouring provinces.

In the Atlantic provinces, with exception of New Brunswick, outflows of workers, on balance, were somewhat higher than inflows. Outflow rates in this area were above the Canada average for all provinces with the exception of Newfoundland. Movements to and from Quebec were relatively small and approximately equal. There was a small net inflow into Ontario, but here also the rate of movement was low. In the west, rates of movement to and from British Columbia were lower than for the Prairie provinces, in which interprovincial movement was relatively high. For the complete three-year period the samples pointed to a small net inflow into Alberta and an outflow from Saskatchewan, in which insured workers were more mobile than in any other province with the possible exception of Prince Edward Island.

Many year-to-year changes in rates of movement were quite marked. It has been noted earlier that local office area movement for Canada held steady between 1956-57 and 1957-58, and then declined in 1958-59. However, interprovincial movement actually increased between 1956-57 and 1957-58, before declining sharply in 1958-59. This suggests that rising unemployment may have contributed to a comparatively large proportion of long moves in search of new jobs. However, in 1958-59 when unemployment continued high, although slightly below the levels of 1957-58, both interprovincial and local office area mobility rates declined along with those recording industrial and occupational changes.

Some of the more noteworthy features of Table 8 of provincial mobility rates are noted briefly. There was a net outflow of insured persons from most provinces when unemployment was rising in 1957-58, with Ontario and Alberta being the only provinces having a substantial net inflow during this period. In 1956-57 and 1958-59, movements to and from most provinces were fairly evenly balanced. For Saskatchewan, outflows predominated in 1956-57 and 1957-58, with an indication of reversal of this tendency in 1958-59 when flows in both directions were comparatively small. In New Brunswick both inflow and outflow rates were high relative to those for other Atlantic provinces in all three years, and, as already noted, they were approximately even in all three years. Although Alberta attracted more insured workers than it lost over the three-year period, outflow rates increased steadily.

More than one-half of insured persons moving out of the Atlantic provinces went to Ontario and Quebec. Practically all of the remainder moved to other provinces within the Atlantic area, and only a very small number moved west of Ontario. The return flow from the central provinces to the Atlantic region was less than one-half of the outflow.

Movement out of Quebec extended east and west to Newfoundland and British Columbia, but about three-quarters of those moving from Quebec went to Ontario. The reverse flow into Quebec from Ontario was slightly smaller.

Outflows from Ontario were more widely distributed than for Quebec. The western provinces ran a close second to Quebec as the destination of insured persons leaving Ontario. The movement from western Canada to Ontario was also fairly large, particularly from Manitoba and British Columbia.

TABLE 8

Provincial Rates of Inflow and Outflow among Insured Workers, 1956-59

Province	Year	Inflow	Outflow	Province	Year	Inflow	Outflow
Newfoundland	1956-57	2	5	Ontario	1956-57	3	3
	1957-58	2	3		1957-58	5	4
	1958-59	3	3		1958-59	2	2
Pr. Edward Island	1956-57	13	12	Manitoba	1956-57	9	8
	1957-58	5	14		1957-58	7	10
	1958-59	4	4		1958-59	7	4
Nova Scotia	1956-57	4	7	Saskatchewan	1956-57	4	16
	1957-58	4	12		1957-58	9	15
	1958-59	4	3		1958-59	6	4
New Brunswick	1956-57	7	7	Alberta	1956-57	7	6
	1957-58	12	10		1957-58	15	8
	1958-59	5	6		1958-59	4	9
Quebec	1956-57	3	3	British Columbia	1956-57	5	4
	1957-58	4	4		1957-58	4	8
	1958-59	2	2		1958-59	4	2

Interprovincial movement within the four western provinces was slightly larger than the outflow to the east. Substantial numbers moved into Ontario from the west, but comparatively few went on to Quebec and the Atlantic provinces.

d) Intraprovincial Mobility. Since interprovincial rates of local office area mobility represent total movement out of local offices for each province, the differences between these rates and rates of provincial outflow represents intraprovincial movement of insured workers.

Unlike interprovincial outflow rates, most intraprovincial rates declined steadily from 1956-57 to 1958-59. It has already been pointed out that interprovincial rates for 1957-58 were almost invariably higher than for 1956-57. This suggests a tendency towards fairly long moves in the early stages of rising unemployment, with comparatively fewer changes at later stages.

Intraprovincial movement in Quebec, Ontario, and British Columbia was generally greater than interprovincial movement. For other provinces it tended to be lower. The possibilities of intraprovincial movement ore generally more limited in the less populous provinces which have comparatively small numbers of local office areas (see Tables 8 and 9).

TABLE 9

Intraprovincial Mobility Rates

Province	1956-57	1957-58	1958-59
Newfoundland	3	2	*
Prince Edward Island	4	3	*
Nova Scotia	7	5	3
New Brunswick	9	3	1
Quebec	7	5	3
Ontario	7	6	5
Manitoba	2	2	1
Saskatchewan	3	5	5
Alberta	7	3	4
British Columbia	10	6	5

* No estimate.

UNEMPLOYMENT AS A FACTOR IN LABOUR MOBILITY

As pointed out in the introduction, no clear-cut distinction can be made between the employed and the unemployed from unemployment insurance book renewal records except at one point in time each year. Although insured persons then can be separated into two groups representing persons who are working and those who are not, nothing is known of the amount of unemployment experienced during the intervening year by the working group, nor of the amount of employment secured by those who were not working.

In spite of this, it seemed worth while to experiment with records for the two groups which, for convenience, will be referred to as the employed and the unemployed. Actually, two experiments were carried out.

The first was designed for a study of total mobility. Since the unemployed have no current industrial or occupational attachment, indications of industrial and occupational change could only be obtained by relating mobility rates to a period of time in which workers had been employed at the beginning, and again at the end, with a record of unemployment coming in between. Two such possibilities existed, the first based upon records of persons unemployed in 1957, but employed in 1956 and again in 1958, and the second including persons who were unemployed in 1958, but employed in 1957 and 1959. These two groups were compared with records of persons with no record of unemployment in the same periods, i.e., 1956-58 and 1957-59. The purpose of this exercise was to establish comparative mobility rates for persons with a record of employment broken by unemployment, and other persons with no record of unemployment.

This experiment, of course, could not provide a representative selection of the unemployed in 1957 and 1958, since substantial proportions of the unemployed totals for these two years did not have a record of employment at preceding and following book renewal dates. Furthermore, mobility rates in this experiment, of necessity, were related to dates two years apart, and consequently could not be compared with other rates in this study which were based upon a one-year period.

The second experiment provided annual local office area mobility rates for complete cross-sections of the employed and unemployed. The same time intervals were used, i.e., 1956-58 and 1957-59, but in this case the unemployed groups included all persons unemployed in 1957 and 1958, some of whom were also unemployed in adjacent years. This meant that some records appeared in both groups of the unemployed, but this was necessary in order to give proper representation to "repeater" unemployment in both comparisons. Proportions of "repeater" unemployment in the two periods were as shown in Table 10. The slightly higher

TABLE 10

Distribution of the Insured Unemployed at Book
Renewal Time
according to Number of Times Unemployed

	1956-58	1957-59
	%	%
Once	85	77
Twice	14	21
Three times	1	2

proportion of "repeaters" in the 1957-59 period may be related to the fact that rates of insured unemployment were higher in 1958 and 1959 than they were in 1956 and 1957. Insured unemployment rates in the four book renewal periods involved in the study were: May, 1956, 7.4; May, 1957, 7.3; May, 1958, 10.3; and May, 1959, 9.8.

Unemployment as a Factor in Total Mobility

Total mobility rates for persons with records of employment broken by unemployment were 78 for 1956-58 and 81 for 1957-59. Corresponding rates for persons with no record of unemployment were 59 and 54. Thus, about four out of five persons with records of broken employment changed jobs at least once in two years, while those always employed at book renewal time changed jobs in less than three out of five cases. If the latter group could have been restricted to persons with absolutely no break in employment, the contrast would undoubtedly have been more marked.

Rates point to higher mobility among the unemployed than among the employed, but for the unemployed group rates increased in the later stages of high unemployment while for the employed they declined (see Table 11). Rates for

TABLE 11

Canada Mobility Rates for the Employed and Unemployed

Mobility measure	Employed		Unemployed	
	1956-58	1957-59	1956-58	1957-59
Total	59	54	78	81
Local office area	15	11	28	28
Industry	30	25	58	63
Occupation	45	41	55	58

all types of mobility were substantially higher for the unemployed than for the employed, with the greatest differences occurring in local office area and industry changes. The contrasts in industry and occupation rates for unemployed and employed are of particular interest. Among the employed, occupation rates were higher than industry rates. Movement within this group will be largely voluntary in character, and high occupational rates may be expected for persons changing to more skilled and better-paid work. On the other hand, among the unemployed industrial mobility was higher than occupational mobility. Persons in this group are more likely to search for the same kind of work as they had before becoming unemployed. Nevertheless, occupational rates for the unemployed were higher than those for the employed.

Unemployment as a Factor in Local Office Area Mobility

From the more comprehensive samples used to examine the influence of unemployment upon area mobility, it was possible to produce local office area rates for the provinces. As in the preceding comparison, local office rates for insured workers with records of unemployment were approximately double those for workers with no record of unemployment.

Behaviour of rates for the two large central provinces were of special interest. Particularly in Quebec and Ontario, mobility among the unemployed appeared to be high relative to mobility among the employed. Rates for workers with a record of unemployment in these provinces were more than double those for workers with no unemployment records, while in other provinces, differences were much less. Possibilities of obtaining employment in Quebec and Ontario by making short moves were undoubtedly greater than elsewhere in Canada.

It is also of interest that Canada rates declined less between 1956-58 and 1957-59 for insured workers with an unemployment record than for those with none. Provincial rates for the former group held steady or increased between 1956-58 and 1957-59 for Newfoundland, New Brunswick, Saskatchewan, and British Columbia (see Table 12).

TABLE 12
Local Office Area Mobility Rates for Insured Workers

Province	With no unemployment record		With an unemployment record	
	1956-58	1957-59	1956-58	1957-59
Newfoundland	6	4	11	12
Prince Edward Island	17	9	25	16
Nova Scotia	16	12	20	19
New Brunswick	14	10	19	21
Quebec	10	7	17	16
Ontario	10	8	23	18
Manitoba	11	8	14	11
Saskatchewan	20	14	32	35
Alberta	12	12	19	18
British Columbia	14	10	25	27
Canada	11	9	20	19

APPENDIX A

Labour Mobility in Canada, 1952-1959

The present study is the second one which has utilized unemployment insurance book renewal data. Some time ago special tabulations were made covering the years 1952-1956.[1] For that period, total mobility rates for Canada were calculated for the insured population including both the employed and the unemployed. These rates were: 1952-53, 54; 1953-54, 54; 1954-55, 56; and 1955-56, 59.

Rates for the next three years representing only those persons with jobs at book renewal time were: 1956-57, 54; 1957-58, 50; and 1958-59, 41.

Experiments with 1956-59 data indicate that exclusion of the unemployed would not have reduced any of the 1952-56 figures by more than three percentage points. It is likely that in 1955-56 labour mobility reached its highest level in the past decade, and that the sharp drop in 1958-59 brought mobility rates to their lowest levels in at least ten years. This statement is supported by hiring and separation rates which were about on a par in the early 1950's and in 1956.

In the 1952-56 study, a change from the ranks of the employed to the unemployed was regarded as a form of involuntary mobility. There was criticism of this concept, and most tabulations for the 1956-59 period excluded records for persons unemployed at unemployment insurance book renewal time.

Evidence from special tabulations for this period suggests very strongly that mobility of those not touched by unemployment is substantially lower than among those who have been. Accordingly in any effort to measure total mobility, including both voluntary and involuntary change of jobs, will be inaccurate if it does not include the unemployed.

---

[1] "A Study of Labour Mobility Based on Unemployment Insurance Records," *Canadian Statistical Review*, July, 1960, p. i.

APPENDIX B

Age, Sex, and Marital Status as Factors in Regional Mobility

Although labour mobility does vary according to age, sex, and marital status, the insured population of Canada is sufficiently homogeneous from province to province that these factors do not contribute materially to interprovincial differences in labour mobility. This can be observed from the comparison shown in Table B 1 of standardized and actual rates of total mobility for the provinces. The standardized rates were computed by using Canada distributions for age, sex, and marital status to compute rates for each province. Tests have indicated that none of the recorded differences were significant.

TABLE B1

Actual and Standardized Provincial Total Labour Mobility Rates, 1956-57

| | | Rates standardized for | | |
Province	Actual rate	Sex	Marital status	Age
Newfoundland	50	50	50	49
Prince Edward Island	61	61	62	59
Nova Scotia	54	53	54	55
New Brunswick	59	59	60	59
Quebec	56	55	55	55
Ontario	51	51	52	51
Manitoba	54	54	55	54
Saskatchewan	59	59	59	59
Alberta	58	58	58	57
British Columbia	57	57	58	58

Since year-to-year differences in mobility rates according to sex, marital status, and age do not vary significantly, illustrative rates in Table B 2 are based only on data for 1956-57. They show that men changed jobs slightly more often than women, and single persons were somewhat more mobile than married workers. The residual marital status group, including a high proportion of heads of "broken" families, was still less mobile. Mobility dropped sharply up to age of thirty-five and more gradually in later years.

TABLE B2

Canada Total Mobility Rates by Sex, Marital Status and Age,
1956-57

Total Mobility	54	Age	
Sex		Under 20	65
Men	56	20-24	60
Women	47	25-34	54
Marital Status		35-44	52
Single	57	45-54	48
Married	51	55-64	47
Other	46	65 and over	46

Discussion

SYLVIA OSTRY

THIS PAPER—or rather its predecessor—represents the first attempt to undertake a wide-ranging survey of mobility in Canada. Far more comprehensive in coverage than any similar effort in the United States or elsewhere, it describes the extent and characteristics of mobility under changing economic conditons, stressing regional aspects, and also deals, although very briefly, with some of the major correlates of mobility—age, sex, and marital status. The authors have on the whole refrained from engaging in speculation and theorizing. The empirical findings are too extensive for me to attempt here a full-scale interpretation—were I capable of doing so. However, for purposes of discussion I should like to raise one or two issues of interpretation on matters which I believe may be of general interest. I apologize in advance if I have neglected other points which appear either to the authors or the audience as clearly of greater significance than those I have chosen to discuss. Before dealing with these issues I would like to comment briefly on the total mobility measure used in the paper.

Measure

The total mobility rate, depending as it does upon the basic units of classification, should, I believe, be labelled "handle with care." Inevitably it will be used to make comparisons with the United States or other countries (that was certainly my first reaction), comparisons which are likely to be very misleading. While many of the American studies have measured occupational and industrial shifts in terms of the Census Bureau's three digit codes, the area definitions varied widely. Moreover, the exclusion in this study of purely inter-employer shifts must also be considered. In the most comprehensive American study—the Six City project—it was found that 20 per cent of all moves involved only movement among firms without any change in occupation or industry. Simple inter-employer moves are more important for some groups than for others; for example, they would be very important for the retail trade, skilled construction workers, skilled textile workers, and garment workers. Since this study has not presented data on mobility rates by industry or occupation the problem of distortion of relationships among industrial and occupational groups does not arise. But evidence suggests that omission of simple inter-employer shifts may have rather substantially reduced the total mobility rates presented and the extent of deflation would not necessarily be the same each year

C.P.S.A. Conference on Statistics, 1961, *Papers*. Printed in the Netherlands.

Interpretation

Now for issues of interpretation. There are two I should like to raise for discussion: one concerns some aspects of the findings on interprovincial comparisons and, briefly, a question about the comparison of the mobility of the employed and the unemployed.

a) Interprovincial Comparisons. The study reveals that the interprovincial dispersion of area mobility rates is greater than that of industrial or occupational rates. Should we expect this? There are a number of related points which might be considered. First regarding area rates. In an urban labour market there are a variety of alternative job opportunities available and changing jobs may involve only changing employers or it may also, and more frequently does, entail a complex move involving employer, industry, and or occupation. In a small town or rural labour market the choice is much more limited and more often a voluntary move in search of a job necessitates leaving the area. Hence the economic geography of the province should be one important determinant of both the extent and nature of mobility. Since there are substantial variations in economic geography among the Canadian provinces one would expect to find substantial variations in area mobility. Another conclusion seems to follow. In provinces where small towns or rural labour markets predominate or are of more than average importance one should find that area mobility is relatively greater than industrial or occupational movement. Here the problem of relating apples and oranges arises. If we look at Table 7, however, where provincial rates are expressed as a ratio to the national rate of 1.0, we can compare across year by year but not down the columns of area rates for any year. The picture is rather muddy. In Prince Edward Island, Nova Scotia, New Brunswick, Saskatchewan and British Columbia, area rates are, in general, relatively higher than other rates, although in 1958-59 Nova Scotia and British Columbia are exceptions. In Ontario, with the most concentrated competitive labour market structure there is little to choose among the three rates. Quebec, with a more dispersed structure, heavily dominated by Montreal, however, also displays no consistent pattern. Alberta and Manitoba appear to contradict the hypothesis. But the most puzzling case in Newfoundland where precisely the reverse situation obtains—area rates are relatively lower than the others. Newfoundland has three labour market areas: Grand Falls, Corner Brook, and St. John's. Transportation facilities are such that movement from one to the other is by no means easy. American studies have shown that the time and cost of travelling are usually more important than distance itself in influencing movement. Whatever the case may be, it seems clear that the Newfoundland example illustrates a general point. One would have to have much fuller information on the nature of the labour market divisions in each of the provinces in order to understand the meaning of the local area mobility rates. The boundaries of the National Employment Service areas are in part determined by administrative considerations relating to the operation of the Unemployment Insurance Commission. The areas do not necessarily coincide with the economists' concept of a labour market, whatever that is. (There is widespread disagreement as to the definition of a local labour market and at least one prominent labour expert has denied its existence in a spatial sense.) I do not want to enter into that argument here, although I should like to return briefly to the question of labour market characteristics at the end of my talk. At this juncture I just want to emphasize that the measure of area mobility in this study is an extraordinarily difficult one to interpret in the absence of more detailed information on the unit of classification. Of course the provinces themselves are not entirely appropriate units for the purpose of economic analysis, and this makes the problem more difficult. It might have been better to have included both inter- and intra-provincial rates in the provincial mobility comparison tables.

Although one would expect to find substantial interprovincial dispersion of area rates, why did industry and occupation rates vary relatively little? Of course words like "substantial" and "relatively little" are impossible to define in this context. What sort of interprovincial variation in, say, industry mobility is to be expected? The authors comment, "Apparently provincial differences in industrial structure have less to do with differences in labour mobility than other factors, such as geography and economic conditions." If I detect an element of surprise in this statement I am inclined to share it. First of all, the industrial composition of employment varies widely across the country. Secondly, the forerunner of this present study showed that industry mobility rates varied considerably ranging from a high in agriculture, fishing, and trapping that was almost three times that in finance and utilities and durable goods manufacturing. Less, but still quite considerable variation was exhibited by occupational mobility rates—the unskilled, for example, being double that of the skilled and semi-skilled group and almost double that of the clerical and sales group. From these facts one would expect Maritime industry and occupation rates to be higher than those of, say, Ontario. Industry rates are clearly higher in the three Maritime provinces in the first two years but not in 1958-59. Occupational rates, however, are generally about the same or lower in the Maritimes than in Ontario. Newfoundland remains an enigma. It would be most interesting to hear the audience's view of these findings. The situation regarding industry rates in 1958-59 seems to me to be related to differing effects of unemployment on different provinces. The Maritimes, much harder hit by unemployment than Ontario, all experienced a sharp decline in industry rates by 1958-59 whereas Ontario rates fell only moderately. This divergent movement produced the results noted above—the elimination of the gap in rates between the Maritimes and Ontario. In general, it would be interesting to estimate the degree of association, if any, between the "sensitivity" of mobility rates and levels of unemployment. Because we are considering at one and the same time in this study structural or long-run determinants of mobility and the short-run effects of changing economic conditions and because our short-run is so short, it becomes very difficult to disentangle interprovincial relationships.

One other aspect of structural differences in labour markets which seems to me worth mentioning concerns the relationship between wage dispersion and mobility. All other things being equal it is reasonable to assume that the higher the degree of uniformity of wages within a given area the lesser the degree of voluntary mobility. The extent of wage dispersion in Ontario is, for example, far less than in the Maritimes and also less than in Quebec. The same relationship generally holds true for mobility rates. One of the factors which strongly influences the extent of wage dispersion are the institutional arrangements of wage determination. Again, all other things being equal, the spread of unionism leads to greater uniformity in wages and working conditions. It has been suggested that the spread of unionism also tends to reduce voluntary mobility. Until more intensive analysis is undertaken one cannot determine whether any or all of these factors are of importance in explaining the findings on interprovincial comparisons of mobility rates. However, it might not be inappropriate to explore some of these points in our discussion today.

b) Unemployment and Mobility. Before concluding I would like briefly to discuss one aspect of the findings comparing mobility rates for the employed and unemployed. The authors state: "The contrasts in industry and occupation rates for unemployed and employed are of particular interest. Among the employed, occupation rates were higher than industry rates. Movement within this group will be largely voluntary in character, and high occupational rates may be expected for persons changing to more skilled and better-paid work. On the other hand, among the unemployed industry mobility was higher than occupational mobility. Persons in this group are more likely to search for the same kind of work as they had before becoming unemployed."

The explanation is plausible, but I think additional comment is required. First, to what extent does voluntarily mobility involve changing to more skilled and better paid work? We know from U.S. labour market studies that, and I quote Lloyd Reynolds, "even among workers who quit their jobs in order to better themselves, a substantial proportion fail to better themselves in terms of wages ... The evidence suggests that a good deal of labour mobility is toward poorer jobs— backward or sideways movement rather than forward movement."[1] Parnes observes that "there is mounting evidence that most voluntary separations by manual workers are made before the worker has 'lined up' a new job" and adds "This in itself seems significant because it means that only a minority of voluntary separations occur because the worker has been 'attracted into' a better job."[2] We do not know very much about the nature of the decisions which prompt job changes and what we do know suggests, at least for the manual worker, a rather haphazard selection process in which personal factors rather than wages or occupational status often play a major role. Granted, however, that a large proportion of voluntary job moves involve crossing occupational boundaries in an upward direction, would not a substantial proportion of involuntary job moves involve crossing occupational boundaries in a downward direction? This would certainly be true if there had been a prolonged period of unemployment.

Secondly, the occupational and industrial structure of the employed and unemployed group would undoubtedly be somewhat dissimilar. One of the most fruitful contributions to labour market theory has been that of Clark Kerr who in the article "The Balkanisation of the Labour Market"[3] stressed that the characteristic patterns of mobility vary widely among differing occupational and industrial groups when markets are institutionalized or, a newer term, "structured". Thus, for example, movement is primarily horizontal in the craft market. For many production jobs horizontal movement is greatly discouraged (because of loss of

---

[1] Reynolds, *Labor Economics and Labor Relations* (Englewood Cliffs, N.J., 1959), p. 398.

[2] Herbert S. Parnes, "The Labor Force and Labor Markets," in *Employment Relations Research,* Industrial Relations Research Association, no. 23 (New York, 1960), p. 28.

[3] In *Labor Mobility and Economic Opportunity,* Social Science Research Council (New York, 1954).

seniority) but vertical intra-plant movement is almost automatic. In manufacturing plants the only "ports of entry" are often the unskilled jobs for manual labour and lower clerical occupations for white collar workers. "Structureless' labour markets have quite different characteristics; examples of such markets would be farm labour, domestic service, and perhaps moonlighting. The analysis of the characteristics of specific labour markets in the Kerr sense is still an area poorly explored. I mention it only because I believe it underlines the need, in future mobility studies, of dealing in less aggregative terms. This study has clearly shown that the mobility rates and patterns of two groups—the employed and the unemployed—are different. I suggest that we cannot fully explain the differences in behaviour of the employed and unemployed groups or of either separately without exploring first the extent and nature of mobility of specific occupational and industrial groups.

In closing, I would like to say that I hope the present authors, who have made such an important contribution in a totally uncharted field of research in this country, will continue their work and seek to answer some of the questions which their stimulating paper has raised.

The Flow of Migration among the Provinces in Canada, 1951-1961

YOSHIKO KASAHARA

FROM THE DEMOGRAPHIC POINT OF VIEW, migration is a basic component of change in the size and composition of a population. From the sociological point of view, it is a sensitive indicator of a social change or a social problem. Not only does a large-scale movement of people occur as a response to major disturbances to the equilibrium of a society or a community, but it also operates as a compulsive force to generate social and economic changes at both the sending and the receiving end of the movement. Stagnancy in migration, on the other hand, often indicates the presence of deterrents to needed adjustment to the changing conditions of life. More knowledge about migration is, therefore, essential to an adequate understanding of social dynamics. Research on regional aspects of economic growth is also dependent upon a prior ordering of facts about the redistribution of population. Furthermore, migration is one of the most important factors that must enter into consideration for various types of planning for administrative and business purposes.

Despite a growing interest in migration and its ramifications, the progress of research in this field has been hampered by the scarcity of pertinent material. Based on limited observations, migration study has not been able to go beyond suggesting merely *ad hoc* hypotheses. Very few of such hypotheses can be operationalized for rigorous testing. Little has thus been accomplished for constructing an integrated theory which will permit a satisfactory explanation of migration either as a "determinant" or a "consequence" of social and economic change.

Paucity of basic data has been especially pronounced in the area of internal migration. In Canada the only national statistics available on the internal movement of people over the last twenty years have been intercensal estimates of net migration derived by indirect methods. Because of the nature of the data and the assumptions that have to be made in applying the methods, the resultant estimates are subject to an indeterminate margin of error. Figures of net migration, moreover, are merely the balances of cross-currents of movement; they cannot reveal the whole story of the complex process of population redistribution. Though useful as measures of net gains and losses of population due to migration for local areas, they yield no information on the total volume or the direction of movement. The

C.P.S.A. Conference on Statistics, 1961, *Papers*. Printed in the Netherlands.

generalized pattern of migration has to be analysed only indirectly by examining the geographic distribution of net gains and losses. Selectivity of migration and its social and economic consequences can merely be inferred in the broadest outline, if at all.

As a preliminary attempt to fill this gap in our knowledge, this study directs its effort to reconstructing the pattern of interprovincial migration in Canada over the last ten years. One of its ultimate objectives is to provide not only the net balances of interprovincial migration but also the separate measures of the total inward and outward movement of population for each of the provinces. The study also aims at identifying the provinces of origin and destination for the two-way streams of migration. The direction of migration as well as the volume of gross movement, which net migration figures conceal, can thus be assessed. The material should also serve as a better basis than estimates of net migration for an analysis of the areas of attraction to migrants and their supply areas. It is possible, moreover, to examine—at least roughly—the variations in the distance travelled by migrants and the intervening forces of "push and pull" that affect the distance as well as the direction of movement.

As will be discussed below, these measures of migration, too, have their limitations. At the present stage of the work on this subject, moreover, analysis has to be restricted to a part of the intended study. Processing of the data required for estimating age-sex differentials in migration and analysing the relation between the shifts in the regional economic trend and the variations in the flow of migration has not been completed. This paper therefore has to be considered as an interim report on larger-scale research under way. Despite all the limitations associated with the results obtained, however, it is hoped that they will shed some light upon the recent pattern of internal migration in this country and serve as a useful introduction to more intensive analyses of the subject after the 1961 census.

CONCEPT AND MEASUREMENT OF MIGRATION

Before the method of estimation adopted in this study and its results are discussed, examination of the concept of migration is in order. In a wide variety of generally accepted definitions of migration one basic element is common: the recognition of a change in the place of usual residence as the essential criterion of migratory movement. This criterion may be looked upon as a pointer to the concept of migration. Obviously, not all movements are migratory. Routine and repetitive intra-community comings and goings that introduce no fundamental change into the established order of collective life are not migratory. Even inter-community movements are not necessarily migratory. Temporary movements of tourists and itinerant salesmen, for example, are not migratory, since these movers remain integral parts of their home communities even if they are on the move over a wide area and over a considerable length of time. The above consideration implies that migration is to be defined as inter-community movement involving a permanent or semi-permanent change in a context of life conditions.

Spelling out the theoretical implications of the concept of migration, however, does not directly lead to the solution of the problem of how to measure it. If properly measured, a change in usual residence could serve as an adequate index of migration; it could be at least a useful indicator of a break from the established mode of life in a given community. Proper measurement of changes in usual residence over space and time, however, is subject to a number of complications which cannot be eliminated within the limits of data and analytical tools available at present. Any attempt to draw a line of demarcation on the continuum of distance and time in order to distinguish migratory from non-migratory shifts in residence would entail a difficult and more or less arbitrary choice. The problem of measurement, moreover, is intertwined with the problem—just as complex—of how to classify various types of migration. Despite a variety of approaches taken to these problems, there still remains a number of "borderline" cases that cannot be measured or classified with any precision. It should further be noted that even if some adequate criteria of migration might be set up, the problem of collecting the data to meet the criteria would persist.

For all the difficulties involved in practical application, however, the concept of migration, if viewed in operational terms, implies the imposition of some boundary lines to be crossed before a movement will be counted as a case of migration. It also implies a shift in residence over a period of considerable duration. The first implication, combined with the fact that short-distance moves tend to predominate in internal migration, leads to the observation that the number of movements to be counted as migratory will be a function of the size and type of the unit area enclosed by a boundary chosen. From the second implication it follows that the pattern of migration as well as the number of migratory movements will be a function of time. In summary, therefore, "a shift in residence of persons or families across a specified boundary over a substantial duration of time" may be adopted as the operational definition of migration. The choice of a boundary and of the duration of time, of course, will have to be more or less arbitrary.

In this study, crossing of provincial boundary lines is taken as the basic criterion of migration, and changes of residence across such boundaries as recorded each month are treated as cases of migration. Theoretically, this approach is not satisfactory. None of the provinces is a homogeneous unit in terms of its social and economic characteristics. Even the size and shape of the provinces, which will have a definite bearing on the volume of migration, are not uniform. The provinces are chosen as the base unit areas, however, for two major considerations: (a) that the data necessary for estimating the volume of gross movement of population for a given unit area are readily available only on the provincial level; and (b) that the results of this study may eventually contribute to refinement of the method of estimating the component of internal migration in the post-censal estimates of provincial populations. It might be added that the over-all pattern of migration in each province is apparently dependent in a large measure upon the type of prevailing economic activity and the size and nature of dominant industrial areas in it. To the extent that this proposition holds true, even the crude estimates of interprovincial migration would serve as useful indices of economic changes taking place.

It should also be noted that, since migration figures derived in this study are based on monthly reports on interprovincial movements of families receiving family allowances, they are bound to include some temporary movements which, according to the above definition, should not be regarded as migratory. Moreover, the data used conceal multiple movements as well as return movements. Figures of gross movement therefore represent all the moves of all types observed or estimated over a given period rather than the totals of migrant individuals or families, some of whom may have returned to their places of origin or have made more than one move within the period. They should be taken *not* as the numbers of migrant persons or families but as indicators of the total magnitude of movement across the provincial boundaries over a given period. The longer the unit period of time, the greater will be the number of return and multiple movements included in measures of gross movement.

These limitations imposed by the available data inevitably demand a further redefinition of the concept of migration. In this study, therefore, migration is defined as all the movements across provincial boundaries involving changes in residence over an indefinite duration of time.

SOURCE OF DATA AND METHOD OF ESTIMATION

As was stated above, the major source of material used in this study is the monthly record of the interprovincial transfer of family allowances accounts. Family allowances statistics, being produced primarily to serve administrative needs, are subject to serious limitations for analytical purposes. They do provide, however, valuable information upon the movement of families with children under sixteen years of age among the provinces, permitting an analysis of the two-way flow of migration between the sending and the receiving areas. There is reason to believe that reporting on interprovincial changes in residence among the recipients of family allowances is quite reliable and up to date and that only a few temporary or non-migratory movements are included in the record. Furthermore, both the 1951 and the 1956 census statistics on age distribution of Canada's population imply that the coverage by family allowances of children under sixteen is almost complete. A small fraction not receiving family allowances would consist largely of immigrants who had not established one year's residence in Canada and hence would not be eligible for this benefit. Exclusion of such immigrants from estimates of interprovincial migration, even though some of them might have joined the stream of internal migration, would not be likely to introduce serious error into the estimates.

The statistics on "migrant families" are cross-tabulated by provinces of their origin and destination. Theoretically, the balance between the figures of in- and out-migration among the provinces should equal zero for the country as a whole. Considerable discrepancy between the two sets of figures has been observed, however, particularly in the earlier years of the period under study. The time lag of the recording of in-migration behind that of out-migration appears to be the major reason for the discrepancy. For the purpose of this study, therefore, figures of out-migration by province of destination are used as the basis for deriving estimates.

Interprovincial migration of families with children under sixteen for each census year represents the sum of monthly figures of out-migrant families from each province distributed among the provinces of destination. Estimates of movement among children under sixteen were derived by the following two steps. First, the annual average of the number of children per family in receipt of family allowances in a given province was multiplied by the number of out-migrant families from that

province. The total outward movement of children in the given age range from
each province was thus estimated on the assumption that there would be no significant difference in the average number of children per family between migrants
and non-migrants. The estimated out-migrant children from a given province were
then allocated among the provinces of destination in proportion to the corresponding distribution of out-migrant families. In this step it is assumed that the
size of migrant families with children under sixteen does not vary by province
of destination. To the extent that the above assumptions are false, an error is
introduced into the estimates of movement of children under sixteen. It is quite
possible that the estimates obtained overstate the interprovincial migration of
children. The margin of error involved, moreover, is likely to vary among the
provinces. No correction of possible error is attempted at this stage, however,
since no information is available upon the family size differentials in migration
in recent years.

The sum of the figures of in-migrant children for a given province of destination
from all the other areas derived by the above step constitutes the estimate of the
total in-migration of children under sixteen for that province. For both the children
under sixteen and their families the in- and out-migration figures estimated by this
procedure equal zero for the country as a whole. No arbitrary adjustment is required for balancing the in- and out-movement figures obtained.

Estimates of the total inward and outward movement of population to and from
any given province is inevitably subject to a far greater degree of error than estimates of migrant families or children because of the lack of crucial information
needed. Several alternative methods, therefore, are being applied experimentally
to this phase of estimation. In this paper the results from only one of them are
presented.

Total in-migration for a given province was first estimated by dividing the number of in-migrant children under sixteen for that province by the corresponding
ratio of this age group to the total in-migrants estimated from the 1941 census
data on interprovincial migration.[1] This preliminary estimate was then adjusted
for change in the age structure of the base population. The basic idea in this
procedure is that the age ratio of migrant population may be assumed to be more
or less stable over an extended period of time. This assumption was dictated by
the lack of information on the age differentials in migration other than what the
1941 census gave on the period of residence of migrant population by age for
the provinces of residence at the census date. To assume that the pattern of age
differentials in migration for the provinces in this country has remained more or
less unchanged over the last twenty years may well be invalid. This attempt, however, is justifiable at this moment, since by checking the results against the 1961
census data on migration after having made a necessary allowance for the effect
of mortality and emigration, we can obtain vital information on the correction
factors required for adjusting the assumed age differentials in migration.

Migration statistics from the coming census are intended primarily to provide
information on the total volume of migration over the intercensal period from
1956 to 1961. Although information on certain characteristics of migrants will
also be obtained, age differentials in migration cannot be measured precisely; for

[1] Since no data on migration are available for Newfoundland, the mean value of the ratios
for the Maritime provinces is used for this province.

migrants can be related only to their age at the census date and not to their age at the time of their movement. The error involved in using the age distribution of migrants from the 1961 census for inferring the age differentials in migration would remain indeterminate within the range of five years. For the postcensal estimates of interprovincial migration, however, the age distribution of migrants within one year's margin of error from the time of their movement is needed. By testing against the 1961 census results the validity of the assumption made and the consequences of its application for deriving total migration, it is possible to arrive at correction factors to be used for adjusting the original ratios for future use.

In this connection, it should also be noted that what is available from the 1941 census is subject to error of another sort. The minimum period of residence used in tabulating the migration data in it was "within 2 years." Furthermore, the so-called interprovincial migration tabulations include immigrants. In order to estimate the age distribution of internal migrants, immigrants arriving within the seventeen months preceding the census date were subtracted from the figures of "interprovincial" migrants within two years, on the assumption that the majority of these immigrants had not moved from their initial provinces of settlement during this time.

Even from the 1941 census data, the age distribution of out-migration is almost impossible to estimate, since the migration tabulation by place of previous residence and age does not specify the time at which migration took place. These data present merely the accumulated results of movement over an indefinite period of time preceding the census in which a series of out-movement took place at various ages in different proportions. The indeterminateness of age at migration is a particularly serious problem for the older age groups for whom the period covered may possibly extend over nearly the entire life span. The lack of specific time reference thus tends to reduce the value of the age breakdown of out-migrant population considerably, compounding the difficulty of interpreting the data progressively as age advances.

Hence the total in-migrant population estimated by this procedure was distributed among the provinces of origin in proportion to the corresponding distribution of "migrant families." The underlying assumption in this procedure is that the forces of "push and pull' operating among the provinces are more or less the same for families with children under sixteen and for the rest of the migrant population.

The annual estimates thus obtained were then summed for each intercensal period. No allowance was made for the effect of mortality or emigration upon interprovincial migrants. The actual number of "surviving" internal migrants to be enumerated in the census or at any given date, therefore, will be somewhat smaller than the estimates indicate.

RESULTS

Since this study is as yet incomplete, analysis at the moment has to be confined to the most significant features revealed by the available results. In examining the figures of estimated migration in the attached tables, some caution is required. Because of the inadequacy of the basic data and the assumptions made at various stages of estimation to fill the gaps in the data, they may be subject to a fairly large margin of error. They should be taken, therefore, strictly as preliminary figures subject to adjustment and possibly to considerable modification after the validity of the assumptions and the appropriateness of the techniques used are checked against the 1961 census data and other independent sources of information. The general pattern and trend of migratory movement in this country for the last ten years, however, may be inferred even from these tentative figures.

During both the intercensal periods under review, the two most highly urbanized and industrialized provinces of the country and the rapidly developing province of Alberta gained population by migration, whereas most other provinces showed some net losses of population due to migration. Although the absolute net gain by migration was largest for Ontario and British Columbia in both periods, the attractive force of these provinces appears to have been weakened in recent years while the "pull" of Alberta has been reinforced. Among the losing provinces those which sustained heavy losses of population by migration were predominantly agricultural areas where few attractive employment opportunities were offered.

Even when allowance is made for the attraction of pleasant climatic conditions along the West coast, the predominant pulling force appears to come from employment opportunities offered by areas of rapid industrial and urban expansion. In other words the general pattern of migration indicated by net migration figures appears to imply a fairly close correlation between the stage of economic development in a given province and its rate of population growth or decline due to migration. It also implies that a trend in net migration gain or loss observed in a given area is dependent upon economic expansion or contraction in the potentially important sending and receiving areas outside.

If judged from estimates of net migration, however, the movement of Canada's population across the provincial boundaries in recent years appears to have been rather negligible. Even for Ontario and British Columbia where the largest net gains through migration were observed, the average addition per year by migration was less than 1 per cent of their respective base populations. Among some of the losing provinces the relative effect of migration upon their population was considerably greater. The annual net loss due to migration, however, was less than 2 per cent of the base population even in Saskatchewan and New Brunswick, which suffered the greatest net loss due to migration over the entire period of ten years. Some of the provinces, in fact, appear to have been little affected by migration.[2]

In terms of gross movement, the picture of population redistribution in this country assumes an entirely different feature. During each of the two intercensal periods under study, more than 200,000 movements of families with children under sixteen years of age across provincial boundaries were recorded, the annual average around 40,000 during the first period mounting to 45,000 during the following period. The estimated total interprovincial movements of children under sixteen ranged from about 90,000 to nearly 120,000 a year, while the total migration is estimated to have amounted to as much as 350,000 to 500,000 a year. In other words, roughly two out of every hundred persons and about the same proportion

---

[2] The only anomalous case is Yukon and Northwest Territories which showed a small net migration loss of families but a substantial net gain in total migration. The estimated total migration for these areas may merely indicate a gross error involved in estimation. In the following analysis, therefore, these areas are excluded from consideration.

of families with children under sixteen in this country apparently changed their residence across provincial boundaries in the course of one year. In many of the provinces, the ratio of interprovincial migrants to the base population has been much higher. Strictly speaking, of course, these estimates should not be taken as the numbers of migrants, but merely as indicators of the total numbers of movements, since they are likely to include multiple and return movements within a given period. Also, they should not be considered as the numbers of "survivors" of migrants at any given point in time, since no allowance is made for the effect of mortality and emigration upon the population of internal migrants. Even then, it is clear that redistribution of the nation's population over the last ten years has been going on at a much more brisk pace than net migration figures indicate. It should be noted that during the 1951-56 period the migratory movement of Canada's population continued to gain momentum until it reached its peak between 1956 and 1958 and then tended to slow down. The total volume of interprovincial migration during each of the recent three years, however, remained somewhat above the level observed in any of the years during the 1951-56 period.

Of this movement throughout the country each year, well over one-fourth flowed into Ontario. British Columbia, Alberta, and Quebec followed Ontario in the given order as areas of concentration of migrants, dividing among them nearly 40 per cent of the total migration. Although the areas of rapid industrial expansion and favourable economic opportunities attracted a large number of migrants from outside, each of them was also an important supply area of migrants for the other provinces. Well over 20 per cent of the nation's migrant population each year originated from Ontario, while the contribution by Quebec, Alberta, and British Columbia to the other provinces each exceeded 10 per cent of the total migrant population. Because of the heavier inflow of migrants, the balances of in- and out-migration remained in favour of Ontario, British Columbia and Alberta. Quebec, on the other hand, continued to sustain a considerable loss by migration throughout the entire ten-year period, despite a large influx of migrants from outside. In relative terms, moreover, the areas where net migration figures indicated a higher degree of stability than others tended to show considerable fluidity of population movement. In none of the provinces, in fact, was migration in any sense a one-way stream; an influx and a corresponding exodus of population of a more or less comparable size were witnessed in every one. The figures of net migration, therefore, grossly misrepresent the volume of in- and out-movement of population involved in affecting the size and composition of the population in each of the provinces and in potentially introducing factors of social and economic change. In order to evaluate the demographic consequences of population redistribution and also to assess their social and economic correlates, it is necessary to examine the relative effects of the inflow and outflow of migration upon a given local population.

At least the extent of interchange of population taking place between any pair of provinces as well as the relative influence of in-migrants from various parts of the country upon a local population may be assessed by examining the cross-tabulations of migrants by provinces of origin and destination. The migrants from the Maritime provinces tend to converge primarily upon Ontario, whereas those from the Prairies are apparently exposed to the two opposing forces of attraction. The dominance of Ontario in the eastern part of the country as an area of superior and diversified economic opportunity appears to reduce the importance of the distance factor involved in migration. Nova Scotia, which ranks second as a centre of concentration of migrants from the three other provinces in the Maritimes, attracts only half as many as Ontario. Out-migrants from Manitoba, on the other hand, are torn between Ontario and the western provinces, while those from Saskatchewan tend to make their way primarily westward. Interchange of migrants between contiguous provinces is most pronounced between Ontario and Quebec, the former sending over a third of its out-migrants to Quebec, which in return sends more than two-thirds of its out-migrants to Ontario. Alberta and British Columbia also exchange their migrant population to a substantial degree. Flow of migration between the Atlantic provinces and the areas to the west of Ontario, on the other hand, is only negligible.

The results also show that the impact of in-migration from Ontario upon the receiving provinces was dominant in every one of the eastern provinces and also on Manitoba. Particularly important is the weight it has on Quebec, where nearly two-thirds of in-migrants come from Ontario. In the other eastern provinces, too, the ratio that the in-migrants from Ontario hold to the total in-migration has been invaribly by far the highest. In the provinces of Saskatchewan, Alberta, and British Columbia, on the other hand, in-migration was predominantly from the contiguous provinces, although even there the effect of migration from Ontario was fairly large.

Studies made in the past have demonstrated that migration is primarily a function of differential growth rates among various areas and the unbalances between the numbers and economic opportunities. Population in an area where economic resources are meagre tends to spill over into areas where more favourable living and working conditions are offered. This is not to deny the importance of other factors such as the attraction of pleasant climatic conditions, cultural opportunities, and personal "motivational linkages" which intervene between external conditions and migratory movement. But in general the stream of migration tends to flow from areas of inferior economic opportunity to those of superior economic opportunity. The above figures appear to support this hypothesis to some extent. If judged by the figures of gross movement, however, areas of attraction to migrants also proved to be major supply areas of migrants to other areas where economic conditions are likely to be less favourable. The picture of demographic adjustment to the prevailing economic conditions has thus become much less clear-cut than net migration figures suggest.

Selectivities involved in the "push and pull" of migrants among different areas of the country must be far more diversified than can be generalized in broad terms of economic determinism. Intertwined with economic variables must be elements of cultural and institutional, as well as social and motivational, factors which act as means of adjustment as well as hindrances to the forces of population movement.

For all the limitations of the indirect method of analysis adopted in this study, moreover, the figures of migration obtained also suggest the potential effect of migration upon local populations. By implication they point to probable shifts in vital trends which would be likely to have a long-range effect upon the subsequent growth and structure of a given population. Furthermore, the wide variations in the volume and direction of movement among different localities point up immediately the complex interplay among economic, social, and cultural variables in the entire proces of demographic change. Many significant problems of migration implied by the obtained results, however, are left unanswered, calling for intensive research in the future.

TABLE 1a

Gross and Net Movement of Families in Receipt of Family Allowances among the Provinces and Territories, Canada, from June, 1951 to May, 1956

Province	Volume			Rate*		
	In	Out	Net	In	Out	Net
Total	205,832	205,832	—	10.7	10.7	—
Nfld.	2,752	2,997	—245	5.3	5.8	—0.5
P.E.I.	2,323	3,157	—834	17.4	23.7	—6.3
N.S.	13,033	15,720	—2,687	14.1	17.0	—2.9
N.B.	9,544	12,022	—2,478	13.1	16.5	—3.4
Que.	25,454	27,764	—2,310	4.8	5.2	—0.4
Ont.	56,223	47,881	8,342	8.9	7.6	1.3
Man.	17,586	20,757	—3,171	16.2	19.1	—2.9
Sask.	17,396	23,418	—6,022	14.6	19.7	—5.1
Alta	27,467	26,565	902	20.1	19.4	0.7
B.C.	31,885	23,288	8,597	19.7	14.4	5.3
Yukon and N.W.T.	2,169	2,263	—94	53.2	55.5	—2.3

* The base is the number of families in receipt of family allowances reported as of June, 1951.

TABLE 1b

Gross and Net Movement of Families in Receipt of Family Allowances among the Provinces and Territories, Canada, from June, 1951 to May, 1956

Province	June, 1951-May, 1952			June, 1952-May, 1953			June, 1953-May, 1954			June, 1954-May, 1955			June, 1955-May, 1956		
	In	Out	Net	In	Out	Net	In	Out	Net	In	Out	Net	In	Out	Net
Total	38,959	38,959	—	39,901	39,901	—	41,633	41,633	—	41,317	41,317	—	44,022	44,022	—
Nfld.	474	680	—206	530	535	—5	511	539	—28	622	594	28	615	649	—34
P.E.I.	461	623	—162	462	611	—149	474	679	—205	462	653	—191	464	591	—127
N.S.	2,312	2,886	—574	2,473	2,924	—451	2,523	3,231	—708	2,850	3,173	—323	2,875	3,506	—631
N.B.	1,717	2,392	—675	1,729	2,228	—499	1,824	2,478	—654	2,115	2,328	—213	2,159	2,596	—437
Que.	4,749	4,861	—112	4,710	5,435	—725	5,146	5,819	—673	5,273	5,431	—158	5,576	6,218	—642
Ont.	11,144	8,799	2,345	10,395	9,354	1,041	11,618	9,747	1,871	10,915	10,192	723	12,151	9,789	2,362
Man.	3,414	4,065	—651	3,400	4,089	—689	3,461	4,056	—595	3,810	4,038	—228	3,501	4,509	—1,008
Sask.	3,205	4,872	—1,667	3,619	4,523	—904	4,046	4,077	—31	3,286	4,476	—1,190	3,240	5,470	—2,230
Alta.	4,929	4,946	—17	5,752	4,829	923	5,676	5,322	354	5,409	5,415	—6	5,701	6,053	—352
B.C.	6,199	4,443	1,756	6,385	4,907	1,478	5,888	5,235	653	6,105	4,562	1,543	7,308	4,141	3,167
Yukon and N.W.T.	355	392	—37	446	466	—20	466	450	16	470	455	15	432	500	—68

TABLE 2a

Gross and Net Movement of Families in Receipt of Family Allowances among the Provinces and Territories, Canada, from June, 1956 to May, 1961 *

Province	Volume			Rate†		
	In	Out	Net	In	Out	Net
Total	233,911	233,911	—	10.3	10.3	—
Nfld.	3,539	4,259	—720	6.0	7.3	—1.2
P.E.I	2,590	2,728	—138	19.6	20.7	—1.0
N.S.	14,972	18,742	—3,770	15.1	18.9	—3.8
N.B.	13,083	13,284	—201	16.9	17.2	—0.2
Que.	29,788	30,402	—614	4.7	4.8	—0.1
Ont.	61,252	55,089	6.163	7.9	7.1	0.8
Man.	19,617	22,800	—3,183	16.0	18.6	—2.6
Sask.	18,435	24,991	—6,556	14.5	19.7	—5.2
Alta.	33,539	31,393	2,146	19.8	18.6	1.2
B.C.	34,397	27,466	6,931	17.3	13.8	3.5
Yukon and N.W.T.	2,699	2,757	—58	56.6	57.8	—1.2

* The volume of migration for the three months from March to May, 1961, is estimated on the basis of the actual volume of gross movement during the preceding nine months and the seasonal variations in migratory movement to and from each province indicated by family allowance statistics over the 1956-60 period.

† The base is the number of families in receipt of family allowances reported as of June, 1956.

TABLE 2b

Gross and Net Movement of Families in Receipt of Family Allowances among the Provinces and Territories, Canada, from June, 1956 to May, 1961*

Province	June, 1956-May, 1957			June, 1957-May, 1958			June, 1958-May, 1959			June, 1959-May, 1960			June, 1960-May, 1961*		
	In	Out	Net	In	Out	Net	In	Out	Net	In	Out	Net	In	Out	Net
Total	49,010	49,010	—	50,005	50,005	—	44,527	44,527	—	45,036	45,036	—	45,333	45,333	—
Nfld.	638	768	−130	705	859	−154	695	848	−153	687	912	−225	814	872	−58
P.E.I.	475	683	−208	559	589	−30	519	486	33	530	487	43	507	483	24
N.S.	2,725	3,973	−1,248	3,286	4,234	−948	3,136	3,617	−481	2,790	3,486	−696	3,035	3,432	−397
N.B.	2,157	2,452	−295	2,867	2,935	−68	2,865	2,520	345	2,540	2,752	−212	2,654	2,625	29
Que.	5,630	6,175	−545	6,285	6,194	91	5,545	6,008	−463	6,098	6,350	−252	6,230	5,675	555
Ont.	13,152	10,234	2,918	13,063	11,702	1,361	11,649	10,907	742	12,070	10,809	1,261	11,318	11,437	−119
Man.	3,848	5,503	−1,655	4,132	5,056	−924	3,762	3,914	−152	3,915	4,179	−264	3,960	4,148	−188
Sask.	3,884	6,667	−2,783	4,102	5,101	−999	3,586	4,156	−570	3,310	4,591	−1,281	3,553	4,476	−923
Alta.	6,671	7,116	−445	6,876	6,755	121	6,547	5,622	925	6,648	5,778	870	6,797	6,122	675
B.C.	9,345	4,790	4,555	7,599	6,097	1,502	5,728	5,913	−185	5,861	5,149	712	5,864	5,517	347
Yukon and N.W.T.	485	649	−164	531	483	48	495	536	−41	587	543	44	601	546	55

* The volume of migration for the three months from March to May, 1961, is estimated on the basis of the actual volume of gross movement during the preceding nine months and the seasonal variations in migratory movement to and from each province indicated by family allowance statistics over the 1956-50 period.

TABLE 3

Estimated Gross and Net Movement of Children under 16 Years of Age among the Provinces and Territories, Canada, from June, 1951 to May, 1956

Province	June, 1951-May, 1952			June, 1952-May, 1953			June, 1953-May, 1954			June, 1954-May, 1955			June, 1955-May, 1956		
	In	Out	Net	In	Out	Net	In	Out	Net	In	Out	Net	In	Out	Net
Total	87,570	87,570	—	89,430	89,430	—	95,360	95,360	—	95,190	95,190	—	103,070	103,070	—
Nfld.	1,070	1,950	—880	1,220	1,560	—340	1,190	1,580	—390	1,450	1,770	—320	1,450	1,950	—500
P.E.I.	1,080	1,620	—540	1,090	1,610	—520	1,140	1,820	—680	1,120	1,770	—650	1,140	1,610	—470
N.S.	5,360	6,870	—1,510	5,750	7,050	—1,300	5,960	7,850	—1,890	6,740	7,740	—1,000	6,950	8,620	—1,670
N.B.	4,000	6,360	—2,360	4,070	5,990	—1,920	4,310	6,720	—2,410	4,960	6,400	—1,440	5,180	7,220	—2,040
Que.	10,230	13,030	—2,800	10,230	14,510	—4,280	11,310	15,710	—4,400	11,680	14,500	—2,820	12,600	16,660	—4,060
Ont.	27,020	17,860	9,160	25,280	19,180	6,100	28,780	20,270	8,510	26,900	21,400	5,500	30,310	20,850	9,460
Man.	7,320	8,620	—1,300	7,300	8,790	—1,490	7,590	8,840	—1,250	8,470	8,890	—420	7,870	10,060	—2,190
Sask.	6,720	10,960	—4,240	7,490	10,220	—2,730	8,690	9,260	—570	7,170	10,250	—3,080	7,190	12,640	—5,450
Alta.	10,470	10,680	—210	12,350	9,710	2,640	12,270	11,710	560	11,900	12,020	—120	12,730	13,680	—950
B.C.	13,560	8,750	4,810	13,730	9,770	3,960	13,120	10,570	2,550	13,770	9,400	4,370	16,690	8,610	8,080
Yukon and N.W.T.	740	870	—130	920	1,040	—120	1,000	1,030	—30	1,030	1,050	—20	960	1,170	—210

TABLE 4

Estimated Gross and Net Movement of Children under 16 Years of Age among the Provinces and Territories, Canada, from June, 1956 to May, 1961

Province	June, 1956-May, 1957			June, 1957-May, 1958			June, 1958-May, 1959			June, 1959-May, 1960			June, 1960-May, 1961		
	In	Out	Net	In	Out	Net	In	Out	Net	In	Out	Net	In	Out	Net
Total	115,490	115,490	—	118,650	118,650	—	106,460	106,460	—	108,910	108,910	—	109,180	109,180	—
Nfld.	1,520	2,330	—810	1,700	2,630	—930	1,690	2,610	—920	1,680	2,820	—1,140	2,020	2,670	—650
P.E.I.	1,170	1,890	—720	1,360	1,630	—270	1,280	1,350	—70	1,320	1,360	—40	1,250	1,340	—90
N.S.	6,630	9,850	—3,220	8,020	10,580	—2,560	7,720	9,040	—1,320	6,990	8,750	—1,760	7,580	8,460	—880
N.B.	5,220	6,870	—1,650	6,940	8,280	—1,340	6,970	7,130	—160	6,230	7,840	—1,610	6,610	7,260	—650
Que.	12,850	16,550	—3,700	14,530	16,600	—2,070	12,870	16,100	—3,230	14,310	17,020	—2,710	14,800	15,070	—270
Ont.	32,940	22,100	10,840	32,860	25,510	7,350	29,480	24,000	5,480	30,840	24,000	6,840	28,450	25,980	2,470
Man.	8,740	12,380	—3,640	9,470	11,530	—2,060	8,690	9,000	—310	9,140	9,700	—560	9,350	9,650	—300
Sask.	8,670	15,670	—7,000	9,250	12,140	—2,890	8,160	10,020	—1,860	7,650	11,150	—3,500	8,030	10,720	—2,690
Alta.	15,110	16,220	—1,110	15,630	15,540	90	14,980	13,100	1,880	15,480	13,580	1,900	15,960	14,270	1,690
B.C.	21,550	10,110	11,440	17,680	13,050	4,630	13,490	12,770	720	13,890	11,280	2,610	13,690	12,310	1,380
Yukon and N.W.T.	1,090	1,520	—430	1,210	1,160	50	1,130	1,340	—210	1,380	1,410	—30	1,440	1,450	—10

TABLE 5

Estimated Gross and Net Movement of Population among the Provinces and Territories, Canada, from June, 1951 to May, 1956

Province	Volume			Rate*		
	In	Out	Net	In	Out	Net
Total	1,853,690	1,853,690	—	13.2	13.2	—
Nfld.	21,660	35,530	—13,870	6.0	9.8	—3.8
P.E.I.	23,280	33,460	—10,180	23.7	34.0	—10.3
N.S.	125,250	149,610	—24,360	19.5	23.3	—3.8
N.B.	79,610	132,840	—53,230	15.4	25.8	—10.3
Que.	229,580	297,940	—68,360	5.7	7.3	—1.7
Ont.	566,670	390,870	175,800	12.3	8.5	3.8
Man.	148,200	177,340	—29,140	19.1	22.8	—3.8
Sask.	143,560	206,120	—62,560	17.3	24.8	—7.5
Alta.	239,420	219,600	19,820	25.5	23.4	2.1
B.C.	248,750	190,570	58,180	21.3	16.4	5.0
Yukon and N.W.T.	27,710	19,810	7,900	110.4	78.9	31.5

* The base is the population as reported in the 1951 census. No adjustment is made for rounding errors.

TABLE 6

Estimated Gross and Net Movement of Population among the Provinces and Territories, Canada, from June, 1956 to May, 1961

Province	Volume			Rate*		
	In	Out	Net	In	Out	Net
Total	2,062,380	2,062,380	—	12.8	12.8	—
Nfld.	28,260	49,320	—21,060	6.8	11.9	—5.1
P.E.I	25,430	28,230	—2,800	25.6	28.4	—2.8
N.S.	145,310	171,270	—25,960	20.9	24.7	—3.7
N.B.	109,330	143,020	—33,690	19.7	25.8	—6.1
Que.	274,450	302,030	—27,580	5.9	6.5	—0.6
Ont.	582,260	453,930	128,330	10.8	8.4	2.4
Man.	163,270	191,530	—28,260	19.2	22.5	—3.3
Sask.	152,120	215,670	—63,550	17.3	24.5	—7.2
Alta	288,320	258,050	30,270	25.7	23.0	2.7
B.C.	258,670	224,540	34,130	18.5	16.1	2.4
Yukon and N.W.T.	34,960	24,790	10,170	111.0	78.7	32.3

* The base is the population as reported in the 1956 census. No adjustment is made for rounding errors.

TABLE 7

Interprovincial Migration of Families in Receipt of Family Allowances, Canada,
from June, 1951 to May, 1956

Province of origin	Total	Province of destination										
		Nfld.	P.E.I.	N.S.	N.B.	Que.	Ont.	Man.	Sask.	Alta.	B.C.	Yukon and N.W.T.
Total	205,832	2,752	2,323	13,033	9,544	25,454	56,223	17,586	17,396	27,467	31,885	2,169
Nfld.	2,997	—	50	765	208	368	1,352	63	23	87	77	4
P.E.I.	3,157	47	—	874	492	275	1,055	168	41	105	92	8
N.S.	15,720	779	764	—	2,442	1,781	7,430	442	167	460	1,383	72
N.B.	12,022	196	454	2,368	—	2,781	5,210	267	63	241	403	39
Que.	27,764	379	182	1,487	2,191	—	20,056	795	315	971	1,321	67
Ont.	47,881	1,127	652	5,370	3,376	16,814	—	6,488	3,071	4,776	5,793	414
Man.	20,757	60	76	377	244	980	7,500	—	4,101	3,258	4,034	127
Sask.	23,418	24	22	169	72	357	3,549	4,570	—	8,124	6,410	121
Alta.	26,565	58	50	374	186	877	4,606	2,330	5,756	—	11,670	658
B.C.	23,288	75	66	1,186	302	1,130	5,021	2,320	3,707	8,822	—	659
Yukon and N.W.T.	2,263	7	7	63	31	91	444	143	152	623	702	—

TABLE 8

Estimated Interprovincial Migration of Families in Receipt of Family Allowances, Canada, from June, 1956 to May, 1961

Province of origin	Total	Province of destination										
		Nfld.	P.E.I.	N.S.	N.B.	Que.	Ont.	Man.	Sask.	Alta.	B.C.	Yukon and N.W.T.
Total	233,911	3,539	2,590	14,972	13,083	29,788	61,252	19,617	18,435	33,539	34,397	2,699
Nfld.	4,259	—	63	1,028	377	577	1,719	138	58	148	146	5
P.E.I.	2,728	88	—	778	565	174	843	51	25	87	104	13
N.S.	18,742	930	769	—	3,774	2,320	8,241	579	173	529	1,361	66
N.B.	13,284	317	488	2,679	—	3,255	5,324	319	93	330	423	56
Que.	30,402	487	161	1,778	2,519	—	21,483	934	319	1,067	1,524	130
Ont.	55,089	1,308	780	6,296	4,554	19,248	—	7,425	2,875	5,676	6,443	484
Man.	22,800	134	95	552	386	1,241	8,065	—	4,297	3,747	4,108	175
Sask.	24,991	37	36	174	97	340	3,352	4,491	—	10,200	6,092	172
Alta.	31,393	120	100	466	310	1,038	5,593	2,886	6,498	—	13,452	930
B.C.	27,466	109	82	1,131	455	1,460	6,050	2,633	3,939	10,939	—	668
Yukon and N.W.T.	2,757	9	16	90	46	135	582	161	158	816	744	—

TABLE 9

Estimated Total Interprovincial Migration, Canada, from June, 1951 to May, 1956

Province of origin	Total	Province of destination										
		Nfld.	P.E.I.	N.S.	N.B.	Que.	Ont.	Man.	Sask.	Alta.	B.C.	Yukon and N.W.T.
Total	1,853,690	21,660	23,280	125,250	79,610	229,580	566,670	148,200	143,560	239,420	248,750	27,710
Nfld.	35,530	—	620	9,170	2,160	4,440	16,300	710	290	1,000	790	50
P.E.I.	33,460	430	—	9,550	4,590	3,020	11,600	1,730	420	1,130	860	130
N.S.	149,610	6,410	7,780	—	20,940	17,700	73,830	4,110	1,570	4,470	11,760	1,040
N.B.	132,840	1,810	5,180	26,230	—	30,990	58,080	2,790	660	2,620	3,850	630
Que.	297,940	3,440	2,050	16,220	20,770	—	220,110	8,190	3,260	10,420	12,410	1,070
Ont.	390,870	7,960	5,670	45,470	24,830	143,290	—	51,870	24,580	39,800	42,290	5,110
Man.	177,340	440	700	3,350	1,890	8,760	66,740	—	34,420	28,490	30,900	1,650
Sask.	206,120	180	210	1,580	580	3,330	33,170	39,990	—	74,150	51,300	1,630
Alta.	219,600	420	450	3,320	1,430	7,820	41,050	19,500	48,150	—	88,980	8,480
B.C.	190,570	520	550	9,780	2,170	9,380	41,660	18,060	28,880	71,650	—	7,920
Yukon and N.W.T.	19,810	50	70	580	250	850	4,130	1,250	1,330	5,690	5,610	—

TABLE 10

Estimated Total Interprovincial Migration, Canada, from June, 1956 to May, 1961

Province of origin	Total	Nfld.	P.E.I.	N.S.	N.B.	Que.	Ont.	Man.	Sask.	Alta.	B.C.	Yukon and N.W.T.
Total	2,062,380	28,260	25,430	145,310	109,330	274,450	582,260	163,270	152,120	288,320	258,670	34,960
Nfld.	49,320	—	760	12,280	3,980	6,970	19,880	1,530	660	1,720	1,440	100
P.E.I.	28,230	790	—	8,580	5,330	1,890	8,840	490	250	900	950	210
N.S.	171,270	7,630	7,530	—	32,210	23,060	77,260	5,230	1,530	4,890	10,960	970
N.B.	143,020	2,950	5,450	29,850	—	36,010	56,330	3,230	960	3,460	3,890	890
Que.	302,030	4,360	1,750	18,810	23,240	—	216,030	8,890	3,140	10,800	13,070	1,940
Ont.	453,930	9,490	6,930	54,440	34,330	168,550	—	58,870	23,000	46,630	45,650	6,040
Man.	191,530	1,010	850	5,010	3,120	11,300	69,550	—	35,930	32,280	30,230	2,250
Sask.	215,670	290	370	1,610	810	3,270	30,290	38,720	—	91,360	46,660	2,290
Alta.	258,050	900	890	4,230	2,510	9,500	49,150	24,240	54,770	—	99,830	12,030
B.C.	224,540	780	740	9,580	3,420	12,550	49,520	20,600	30,480	88,630	—	8,240
Yukon and N.W.T.	24,790	60	160	920	380	1,350	5,410	1,470	1,400	7,650	5,990	—

TABLE 11

Percentage Distribution of Migrant Families in Receipt of Family Allowances by Province of Destination for the Provinces and Territories of Origin, Canada, from June, 1951 to May, 1956

Province of origin	Total	Province of destination										
		Nfld.	P.E.I.	N.S.	N.B.	Que.	Ont.	Man.	Sask.	Alta.	B.C.	Yukon and N.W.T.
Total	100.0	1.3	1.1	6.3	4.6	12.4	27.3	8.5	8.5	13.3	15.5	1.1
Nfld.	100.0	—	1.7	25.5	6.9	12.3	45.1	2.1	0.8	2.9	2.6	0.1
P.E.I	100.0	1.5	—	27.7	15.6	8.7	33.4	5.3	1.3	3.3	2.9	0.3
N.S.	100.0	5.0	4.9	—	15.5	11.3	47.3	2.8	1.1	2.9	8.8	0.4
N.B.	100.0	1.6	3.8	19.7	—	23.1	43.3	2.2	0.5	2.0	3.4	0.3
Que.	100.0	1.4	0.7	5.4	7.9	—	72.2	2.9	1.1	3.5	4.8	0.2
Ont.	100.0	2.4	1.4	11.2	7.0	35.1	—	13.6	6.4	10.0	12.1	0.9
Man.	100.0	0.3	0.4	1.8	1.2	4.7	36.1	—	19.8	15.7	19.4	0.6
Sask.	100.0	0.1	0.1	0.7	0.3	1.5	15.2	19.5	—	34.7	27.4	0.5
Alta.	100.0	0.2	0.2	1.4	0.7	3.3	17.3	8.8	21.7	—	43.9	2.5
B.C.	100.0	0.3	0.3	5.1	1.3	4.8	21.6	10.0	15.9	37.9	—	2.8
Yukon and N.W.T.	100.0	0.3	0.3	2.8	1.4	4.0	19.6	6.3	6.7	27.5	31.0	—

TABLE 12

Percentage Distribution of Estimated Migrant Families in Receipt of Family Allowances by Province of Destination for the Provinces and Territories of Origin, Canada, from June, 1956 to May, 1961

Province of origin	Total	Province of destination										
		Nfld.	P.E.I.	N.S.	N.B.	Que.	Ont.	Man.	Sask.	Alta.	B.C.	Yukon and N.W.T.
Total	100.0	1.5	1.1	6.4	5.6	12.7	26.2	8.4	7.9	14.3	14.7	1.2
Nfld.	100.0	—	1.5	24.1	8.9	13.5	40.4	3.2	1.4	3.5	3.4	0.1
P.E.I.	100.0	3.2	—	28.5	20.7	6.4	30.9	1.9	0.9	3.2	3.8	0.5
N.S.	100.0	5.0	4.1	—	20.1	12.4	44.0	3.1	0.9	2.8	7.3	0.4
N.B.	100.0	2.4	3.7	20.2	—	24.5	40.1	2.4	0.7	2.5	3.2	0.4
Que.	100 0	1.6	0.5	5.8	8.3	—	70.7	3.1	1.0	3.5	5.0	0.4
Ont.	100.0	2.4	1.4	11.4	8.3	34.9	—	13.5	5.2	10.3	11.7	0.9
Man.	100.0	0.6	0.4	2.4	1.7	5.4	35.4	—	18.8	16.4	18.0	0.8
Sask.	100.0	0.1	0.1	0.7	0.4	1.4	13.4	18.0	—	40.8	24.4	0.7
Alta.	100.0	0.4	0.3	1.5	1.0	3.3	17.8	9.2	20.7	—	42.9	3.0
B.C.	100.0	0.4	0.3	4.1	1.7	5.3	22.0	9.6	14.3	39.8	—	2.4
Yukon and N.W.T.	100.0	0.3	0.6	3.3	1.7	4.9	21.1	5.8	5.7	29.6	27.0	—

TABLE 13

Percentage Distribution of Estimated Migrant Population by Province of Destination for the Provinces and Territories of Origin, Canada, from June, 1951 to May, 1956

Province of origin	Province of destination											
	Total	Nfld.	P.E.I.	N.S.	N.B.	Que.	Ont.	Man.	Sask.	Alta.	B.C.	Yukon and N.W.T.
Total	100.0	1.2	1.3	6.8	4.3	12.4	30.6	8.0	7.7	12.9	13.4	1.5
Nfld.	100.0	—	1.7	25.8	6.1	12.5	45.9	2.0	0.8	2.8	2.2	0.1
P.E.I.	100.0	1.3	—	28.5	13.7	9.0	34.7	5.2	1.3	3.4	2.6	0.4
N.S.	100.0	4.3	5.2	—	14.0	11.8	49.3	2.7	1.0	3.0	7.9	0.7
N.B.	100.0	1.4	3.9	19.7	—	23.3	43.7	2.1	0.5	2.0	2.9	0.5
Que.	100.0	1.2	0.7	5.4	7.0	—	73.9	2.7	1.1	3.5	4.2	0.4
Ont.	100.0	2.0	1.5	11.6	6.4	36.7	—	13.3	6.3	10.2	10.8	1.3
Man.	100.0	0.2	0.4	1.9	1.1	4.9	37.6	—	19.4	16.1	17.4	0.9
Sask.	100.0	0.1	0.1	0.8	0.3	1.6	16.1	19.4	—	36.0	24.9	0.8
Alta.	100.0	0.2	0.2	1.5	0.7	3.6	18.7	8.9	21.9	—	40.5	3.9
B.C.	100.0	0.3	0.3	5.1	1.1	4.9	21.9	9.5	15.2	37.6	—	4.2
Yukon and N.W.T.	100.0	0.3	0.3	2.9	1.3	4.3	20.8	6.3	6.7	28.7	28.3	—

TABLE 14

Percentage Distribution of Estimated Migrant Population by Province of Destination for the Provinces and Territories of Origin, Canada, from June, 1956 to May, 1961

Province of origin	Total	Province of destination										
		Nfld.	P.E.I.	N.S.	N.B.	Que.	Ont.	Man.	Sask.	Alta.	B.C.	Yukon and N.W.T.
Total	100.0	1.4	1.2	7.0	5.3	13.3	28.2	7.9	7.4	14.0	12.5	1.7
Nfld.	100.0	—	1.5	24.9	8.1	14.1	40.3	3.1	1.3	3.5	2.9	0.2
P.E.I.	100.0	2.8	—	30.4	18.9	6.7	31.3	1.7	0.9	3.2	3.4	0.7
N.S.	100.0	4.5	4.4	—	18.8	13.5	45.1	3.0	0.9	2.9	6.4	0.6
N.B.	100.0	2.1	3.8	20.9	—	25.2	39.4	2.3	0.7	2.4	2.7	0.6
Que.	100.0	1.4	0.6	6.2	7.7	—	71.5	3.0	1.0	3.6	4.3	0.6
Ont.	100.0	2.1	1.5	12.0	7.6	37.1	—	13.0	5.1	10.3	10.1	1.3
Man.	100.0	0.5	0.4	2.6	1.6	5.9	36.3	—	18.8	16.9	15.8	1.2
Sask.	100.0	0.1	0.2	0.7	0.4	1.5	14.0	18.0	—	42.4	21.6	1.1
Alta.	100.0	0.3	0.3	1.6	1.0	3.7	19.0	9.4	21.2	—	38.7	4.7
B.C.	100.0	0.3	0.3	4.3	1.5	5.6	22.1	9.2	13.6	39.5	—	3.7
Yukon and N.W.T.	100.0	0.2	0.6	3.7	1.5	5.5	21.8	5.9	5.6	30.9	24.2	—

TABLE 15

Percentage Distribution of Migrant Families in Receipt of Family Allowances by Province of Origin for the Provinces and Territories of Destination, Canada, from June, 1951 to May, 1956

Province of origin		Province of destination										
	Total	Nfld.	P.E.I.	N.S.	N.B.	Que.	Ont.	Man.	Sask.	Alta.	B.C.	Yukon and N.W.T.
Total	100.0	100.0	100.0	100.0	100.0	100.0	100.0	100.0	100.0	100.0	100.0	100.0
Nfld.	1.5	—	2.2	5.9	2.2	1.4	2.4	0.4	0.1	0.3	0.2	0.2
P.E.I.	1.5	1.7	—	6.7	5.1	1.1	1.9	1.0	0.2	0.4	0.3	0.4
N.B.	7.6	28.3	32.9	—	25.6	7.0	13.2	2.5	1.0	1.7	4.3	3.3
N.S.	5.8	7.1	19.5	18.2	—	10.9	9.3	1.5	0.4	0.9	1.3	1.8
Que.	13.5	13.8	7.8	11.4	23.0	—	35.7	4.5	1.8	3.5	4.1	3.1
Ont.	23.3	41.0	28.1	41.2	35.4	66.1	—	36.9	17.7	17.4	18.2	19.1
Man.	10.1	2.2	3.3	2.9	2.6	3.9	13.3	—	23.6	11.9	12.7	5.9
Sask.	11.4	0.9	0.9	1.3	0.8	1.4	6.3	26.0	—	29.6	20.1	5.6
Alta.	12.9	2.1	2.2	2.9	1.9	3.4	8.2	13.2	33.1	—	36.6	30.3
B.C.	11.3	2.7	2.8	9.1	3.2	4.4	8.9	13.2	21.3	32.1	—	30.4
Yukon and N.W.T.	1.1	0.3	0.3	0.5	0.3	0.4	0.8	0.8	0.9	2.3	2.2	—

TABLE 16

Percentage Distribution of Estimated Migrant Families in Receipt of Family Allowances by Province of Origin for the Provinces and Territories of Destination, Canada, from June, 1956 to May, 1961

Province of origin	Total	Province of destination										
		Nfld.	P.E.I.	N.S.	N.B.	Que.	Ont.	Man.	Sask.	Alta.	B.C.	Yukon and N.W.T.
Total	100.0	100.0	100.0	100.0	100.0	100.0	100.0	100.0	100.0	100.0	100.0	100.0
Nfld.	1.8	—	2.4	6.9	2.9	1.9	2.8	0.7	0.3	0.4	0.4	0.2
P.E.I.	1.2	2.5	—	5.2	4.3	0.6	1.4	0.3	0.1	0.3	0.3	0.5
N.S.	8.0	26.3	29.7	—	28.8	7.8	13.5	3.0	0.9	1.6	4.0	2.4
N.B.	5.7	9.0	18.8	17.9	—	10.9	8.7	1.6	0.5	1.0	1.2	2.1
Que.	13.0	13.8	6.2	11.9	19.3	—	35.1	4.8	1.7	3.2	4.4	4.8
Ont.	23.6	37.0	30.1	42.1	34.8	64.6	—	37.8	15.6	16.9	18.7	17.9
Man.	9.7	3.8	3.7	3.7	3.0	4.2	13.2	—	23.3	11.2	11.9	6.5
Sask.	10.7	1.0	1.4	1.2	0.7	1.1	5.5	22.9	—	30.4	17.7	6.4
Alta.	13.4	3.4	3.9	3.1	2.4	3.5	9.1	14.7	35.2	—	39.1	34.5
B.C.	11.7	3.1	3.2	7.6	3.5	4.9	9.9	13.4	21.4	32.6	—	24.7
Yukon and N.W.T.	1.2	0.3	0.6	0.6	0.4	0.5	1.0	0.8	0.9	2.4	2.2	—

TABLE 17

Percentage Distribution of Estimated Migrant Population by Province of Origin for the Provinces and Territories of Destination, Canada, from June, 1951 to May, 1956

Province of origin	Total	Province of destination										
		Nfld.	P.E.I.	N.S.	N.B.	Que.	Ont.	Man.	Sask.	Alta.	B.C.	Yukon and N.W.T.
Total	100.0	100.0	100.0	100.0	100.0	100.0	100.0	100.0	100.0	100.0	100.0	100.0
Nfld.	1.9	—	2.7	7.3	2.7	1.9	2.9	0.5	0.2	0.4	0.3	0.2
P.E.I.	1.8	2.0	—	7.6	5.8	1.3	2.0	1.2	0.3	0.5	0.3	0.5
N.S.	8.1	29.6	33.4	—	26.3	7.7	13.0	2.8	1.1	1.9	4.7	3.7
N.B.	7.2	8.4	22.3	20.9	—	13.5	10.2	1.9	0.5	1.1	1.5	2.3
Que.	16.1	15.9	8.8	13.0	26.1	—	38.8	5.5	2.3	4.4	5.0	3.9
Ont.	21.1	36.7	24.4	36.3	31.2	62.4	—	35.0	17.1	16.6	17.0	18.4
Man.	9.6	2.0	3.0	2.7	2.4	3.8	11.8	—	24.0	11.9	12.4	6.0
Sask.	11.1	0.8	0.9	1.3	0.7	1.5	5.9	27.0	—	31.0	20.6	5.9
Alta.	11.8	1.9	1.9	2.7	1.8	3.4	7.2	13.2	33.5	—	35.8	30.6
B.C.	10.3	2.4	2.4	7.8	2.7	4.1	7.4	12.2	20.1	29.9	—	28.6
Yukon and N.W.T.	1.1	0.2	0.3	0.5	0.3	0.4	0.7	0.8	0.9	2.4	2.3	—

TABLE 18

Percentage Distribution of Estimated Migrant Population by Province of Origin for the Provinces and Territories of Destination, Canada, from June, 1956 to May, 1961

Province of origin	Total	Province of destination										
		Nfld.	P.E.I.	N.S.	N.B.	Que.	Ont.	Man.	Sask.	Alta.	B.C.	Yukon and N.W.T.
Total	100.0	100.0	100.0	100.0	100.0	100.0	100.0	100.0	100.0	100.0	100.0	100.0
Nfld.	2.4	—	3.0	8.5	3.6	2.5	3.4	0.9	0.4	0.6	0.6	0.3
P.E.I.	1.4	2.8	—	5.9	4.9	0.7	1.5	0.3	0.2	0.3	0.4	0.6
N.S.	8.3	27.0	29.6	—	29.5	8.4	13.3	3.2	1.0	1.7	4.2	2.8
N.B.	6.9	10.4	21.4	20.5	—	13.1	9.7	2.0	0.6	1.2	1.5	2.5
Que.	14.6	15.4	6.9	12.9	21.3	—	37.1	5.4	2.1	3.7	5.1	5.5
Ont.	22.0	33.6	27.3	37.5	31.4	61.4	—	36.1	15.1	16.2	17.6	17.3
Man.	9.3	3.6	3.3	3.4	2.9	4.1	11.9	—	23.6	11.2	11.7	6.4
Sask.	10.5	1.0	1.5	1.1	0.7	1.2	5.2	23.7	—	31.7	18.0	6.6
Alta.	12.5	3.2	3.5	2.9	2.3	3.5	8.4	14.8	36.0	—	38.6	34.4
B.C.	10.9	2.8	2.9	6.6	3.1	4.6	8.5	12.6	20.0	30.7	—	23.6
Yukon and N.W.T.	1.2	0.2	0.6	0.6	0.3	0.5	0.9	0.9	0.9	2.7	2.3	—

Population Migration
in the Atlantic Provinces*

KARI LEVITT

IN THIS PAPER we set ourselves the expository task of presenting statistical information on net population movements in the Atlantic provinces derived almost exclusively from one source, the census of population. This exercise, gains significance if it is regarded as providing information essential to any discussion of the chronically slower rate of economic development in the Atlantic Provinces. Furthermore, we believe the findings of this study have certain policy implications.

We concluded that out-migration from the poorer areas of the region has proceeded at very high rates and that the pattern of out-migration on the whole conforms to economic incentives. We further conclude that even the observed heavy movement of people out of these areas will not prevent a growing surplus of man-power in the forseeable future. The current bulge in the labour force in conjunction with declining employment in resource-based industries and the cumulative concentration of secondary manufacturing in central Canada may be expected to result in a widening gap in earned income per person between the Atlantic region and central Canada. Lower labour force participation rates, higher unemployment rates and the waste of manpower in low-productivity subsistence occupations will be a continuing feature of the region despite heavy spontaneous out-migration unless there is deliberate policy intervention to influence the economic structure of the region.

In so far as out-migration is heavily concentrated among young people it results in a highly unfavourable ratio of labour force to population. This factor, together with the retarding effect of a differentially slower population growth rate on the regional market, reflected most clearly in residential construction, accounts for the fact that spontaneous out-migration, especially if it is heavily concentrated among young people, does not constitute a solution to the problems of the region.

Two alternative policy solutions have been suggested: large scale assisted accelerated out-migration or "the creation of a substantial industrial sector"[1] in the Region. We suggest that the characteristics of the labour supply and particularly

* This paper is essentially a summary of the results of a study on *Population Movements in the Atlantic Provinces* commissioned by the Atlantic Provinces Research Board and published by Atlantic Provinces Economic Council (Halifax and Fredericton, 1960).

[1] A. K. Caincross, *Economic Development and the Atlantic Provinces,* sponsored by Atlantic Provinces Research Board (Fredericton, 1961), p. 25.

C.P.S.A. Conference on Statistics, 1961, *Papers.* Printed in the Netherlands.

its observed physical mobility pattern are fundamental data to the policy-maker, in considering whether to "move the people out" or "move the industry in."

A BRIEF NOTE ON METHOD

The main source of information is the census of Canada. The census does not, unfortunately, provide us with direct information on migration, except for the census year of 1941. We are thus reduced to dealing with a fictitious character called a "net migrant." Net migration movements are the result of much larger (and unknown) gross movements in many directions. To illustrate, we may sup-- pose that there are three locations, *A, B,* and *C. A* and *B* each show in-migration of plus 5, while *C* shows out-migration of minus 10. There is absolutely no reason to suppose that half the people leaving *C* go to *A* and half go to *B*. It is quite possible that there is no direct movement from *C* to *A*, but, on the contrary, a movement from *A* to *C*. This warning should be born in mind throughout the subsequent account of net migration. The fiction is, however, very useful. It tells us which areas are gaining, and which are losing population by migration.

Net migration can be estimated by two methods: the *natural increase* method and the *survival method*. In both cases the accuracy of the estimate depends on the quality of Census data. Canada has a *de jure* census and thus the recorded place of residence is sometimes left to the discretion of the enumerator. Floating populations such as men in construction camps and non-permanent members of the Armed Services are enumerated as residing at their normal place of residence, and not at the camp where they are found.

The natural increase method is more accurate, but is only possible when we have data on births and deaths relating to the population we are dealing with. The procedure is based on the following identity:

> Population at census, 1951
> plus Births, 1951-56 (5 years)
> less Deaths, 1951-56 (5 years)
> less Population at Census, 1956
> equals Net Migration, 1951-56

If migration is *into* an area, we define it as positive; if it is *out of* an area, we define it as negative. Details of this method of estimating net migration 1951-56 for the four Atlantic provinces are presented in the "Appendix on Statistical Method" together with an evaluation of accuracy of the procedure.

The survival method is based on life tables. For 1941-51 we used the Maritime rates as given in the Canadian life table. For the 1951-56 period we were forced to use all-Canadian rates as these were the only ones available. Male and female net migrations were estimated separately, using survival rates appropriate for the given age-group and intercensal interval. Estimates of net migration obtained in this manner fell short of total net migration because they excluded migrants under five (or under ten) years of age. In order to show the difference between estimates obtained by this method and estimate obtained by the natural increase method, we added migrants 0-4. A discussion of our reconciliation of results arrived at by the two methods will be found in the "Appendix on Method."

Obviously, we did not need to rely on the cruder survival method to obtain estimates of provincial net migration for recent intercensal intervals. We did however make extensive use of this method in order to obtain the age-specific net migration rates, net migration rates for national, language and religious groups and net migration estimates for the rural farm population. We again refer the reader to the "Appendix on Statistical Method" where we have set out the procedure for some of these operations.

HISTORICAL TREND OF MIGRATION FROM THE ATLANTIC PROVINCES

It is a well-known fact of Canadian economic development that the Maritime provinces have experienced a slower rate of population growth than the rest of Canada in every decade since Confederation. Prince Edward Island actually experienced a population decline between 1891 and 1931, while Nova Scotia's population declined in the decade 1921-31 (see Table 4).

Population growth for the Maritime provinces in the period 1901-1956 was as follows: Prince Edward Island, 4 per cent; Nova Scotia, 51 per cent; New Brunswick 68 per cent. Total Canadian population (excluding Newfoundland) increased by 192 per cent with the following growth rates in the other provinces: Ontario, 148 per cent; Quebec, 181 per cent; Manitoba, 233 per cent; Saskatchewan, 865 per cent; Alberta, 1,438 per cent; and British Columbia, 683 per cent.

Another reflection of the differentially lower population growth rates in the Maritime provinces can be seen in the change in the geographical distribution of Canada's population. The three Maritime Provinces accounted for 16.7 per cent of Canada's total population in 1901 and only 8.5 per cent in 1956.

Population growth in any area is determined by two factors: natural increase and migration. Fertility and mortality rates in the Maritimes are not essentially different from those prevailing elsewhere in Canada. The slower rate of population growth in this region is thus the result of a steady stream of migration out of the area. An additional and related factor is that the Maritime region has received relatively few foreign immigrants and has not been able to retain all of those who landed there. Estimates show a net loss by migration of 586,000 persons from the Maritime region during the period 1881-1956. The rest of Canada, during this same period gained 1,810,000 people by migration (see Table 1). On a net basis, the Maritime loss by migration amounted to nearly half (43 per cent) the population gain in the rest of Canada in the period 1881-1921 and to one-quarter (24 per cent) in the period 1921-56.

Each of the three Maritime provinces experienced net out-migration in every decade since 1881.[2] There is only one exception to this statement. Nova Scotia showed net in-migration between 1931 and 1941. The reasons are to be found in the special conditions created by the depression where Maritimers found they could at least subsist and survive at home.

While we have no net migration estimates for the period 1871-81, population growth rates in the Maritimes during this decade were only slightly lower than

---

[2] Estimates of provincial net migration cannot be made before 1881-91.

TABLE 1

Net Migration Estimates for Canada and the Provinces, 1891-1921

(in thousands)

	1881-91	1891-1901	1901-11	1911-21	1921-31	1931-41	1941-51	1951-56	1881-1956
Prince Edward Island	−14	−17	−17	−14	−9	−3	−12	−8	−94
Nova Scotia	−43	−40	−28	−37	−62	8	−39	−11	−252
New Brunswick	−44	−32	−30	−25	−36	−10	−42	−21	−240
Quebec	−132	−121	−29	−99	−10	−2	−12	98	−306
Ontario	−84	−144	74	46	129	77	305	375	+778
Manitoba	52	48	111	24	−10	−48	−61	—	+116
Saskatchewan }			283	78	−5	−158	−200	−37	+355
Alberta	21*	68*	218	85	22	−42	−7	64	+355
British Columbia	37	58	164	58	101	82	231	135	+866
Canada	−205*	−181*	746	117	120	−96	163	595	1224
Maritime Region	−101	−89	−75	−76	−107	−5	−93	−40	−586
Rest of Canada	−104	−92	821	193	227	−91	256	635	+1810

Note: These estimates refer to migrants aged ten and over.

* Northwest Territories and Yukon included from 1881 to 1901 and excluded thereafter.

Source: For period, 1881-1921, Nathan Keyfitz "The Growth of Canadian Population," *Population Studies*, vol. IV, no. 1, June, 1950. For period 1921-31 we made our own estimates for the Maritime Procinces based on the more accurate national increase method using available vital statistics; Keyfitz estimates based on the survival method are shown for the other provinces. For period 1931-1951 see *Canada Year Book, 1959*, p. 163. For period 1951-1956, we made our own estimates, see Appendix on Statistical Method.

in Canada. The population of Prince Edward Island increased by 16 per cent and that of Nova Scotia by 14 per cent in the period 1871-81. Neither of these provinces again experienced such rapid growth. The relative decline of the Maritime economy reflected in heavy out-migration and almost stationary population became evident in the last two decades of the nineteenth century. It should be pointed out, however, that a similar condition of stagnation was found in central Canada at this time. Out-migration from Ontario and Quebec greatly exceeded that from the Maritimes in volume. The first decade of the twentieth century, however, reversed this trend for Ontario which has attracted large net in-migrations ever since. In the period 1881-1921 out-migration from Quebec alone exceeded that of the entire Maritime region. The Prairie provinces were, of course, an area of large in-migration until 1921. The subsequent heavy movement of people out of the Prairie provinces has now ceased, with the exception of Saskatchewan which continued to show net out-migration between 1951 and 1956. British Columbia has been a province of in-migration in every decade since its formation.

Further examination of Table 1 indicates that Nova Scotia shared some of the benefits of the "wheat boom." Net out-migration for Nova Scotia in the period 1901-11 slowed considerably. Nova Scotia seems to have received and retained, at least during this period, some of the immigrants pouring into Canada from Europe. New Brunswick and Prince Edward Island, however, continued to export population at the same rates as prevailed before 1901. In the period following the First World War net out-migration from Nova Scotia was exceptionally high, exceeding the rate of natural increase and resulting in an actual decline in population. The depression slowed net out-migration in New Brunswick and Prince Edward Island and reversed the direction of migration in Nova Scotia for the 1931-41 period. The pattern of net out-migration was resumed after the Second World War.

Net migration as a percentage of natural increase in any area is an indicator of the rate at which the natural population increase is leaving the area. If this rate exceeds 100 per cent there will be an absolute decline in population. Table 2 shows population, natural increase and net migration for the period 1921-1956 in detail for each Atlantic province, and for males and females separately.

We can thus observe that Prince Edward Island had the highest rate of net out-migration (75 per cent of natural increase over the period 1921-56) while Nova Scotia (38 per cent) and New Brunswick (39 per cent) experienced only half that rate of out-migration. Newfoundland, despite its very high birth rate and low incomes, had the lowest rate of out-migration of the region (22 per cent). The Maritime region has lost 41 per cent of its natural increase since 1921, while the same loss for the Atlantic region is 37 per cent.

If we examine trends we may observe that the net migration rate out of Newfoundland has been falling steadily and that, in the 1951-56 period, there appears to be for the first time in modern history a net inflow of population to Newfoundland. Net out-migration from Nova Scotia slowed from 38 per cent of natural increase in 1941-51 to 17 per cent in 1951-56. A similar but much less pronounced trend is observable in New Brunswick (45 per cent and 35 per cent). Prince Edward Island continued to sustain out-migration at very high rates throughout the period.

This cursory examination of trends is a warning against crudely projecting past

TABLE 2

Population, Natural Increase, and Net Migration, 1921-1956,
(in thousands)

Year	Population	Period	Natural increase	Net migration	Net out-migration as percentage of natural increase per cent
			Prince Edward Island		
Males					
1921	44.9				
1931	45.4	1921-31	4.2	—3.7	88
1941	49.2	1931-41	5.0	—1.1	23
1951	50.2	1941-51	7.9	—6.9	87
1956	50.5	1951-56	4.4	—4.1	93
		1921-56	21.5	—15.8	73
Females					
1921	43.7				
1931	42.6	1921-31	4.0	—5.1	127
1941	45.8	1931-41	4.7	—1.5	33
1951	48.2	1941-51	7.9	—5.5	70
1956	48.8	1951-56	4.6	—4.0	87
		1921-56	21.2	—16.1	76
Total					
1921	88.6				
1931	88.0	1921-31	8.2	—8.8	107
1941	95.0	1931-41	9.7	—2.6	27
1951	98.4	1941-51	15.8	—12.4	78
1956	99.3	1951-56	9.0	—8.1	90
		1921-56	42.7	—31.9	75
			Nova Scotia		
Males					
1921	266.5				
1931	263.1	1921-31	25.8	—29.2	113
1941	296.0	1931-41	28.4	4.6	—16
1951	325.0	1941-51	51.1	—22.1	46
1956	353.2	1951-56	31.1	2.9	8
		1921-56	136.4	—49.6	36
Females					
1921	257.4				
1931	249.7	1921-31	25.4	—33.1	130
1941	281.9	1931-41	29.0	3.2	—11
1951	317.6	1941-51	52.4	—16.8	32
1956	341.5	1951-56	32.0	—8.1	24
		1921-56	138.8	—54.8	39
Total					
1921	523.8				
1931	512.8	1921-31	51.3	—62.3	121
1941	578.0	1931-41	57.4	7.8	—14
1951	642.6	1941-51	103.5	—38.9	38
1956	694.7	1951-56	63.1	—11.0	17
		1921-56	275.3	—104.4	38

TABLE 2 (*continued*)

Year	Population	Period	Natural increase	Net migration	Net out-migration as percentage of natural increase per cent
			New Brunswick		
Males					
1921	197.4				
1931	208.6	1921-31	29.5	—17.2	60
1941	234.1	1931-41	29.3	—3.9	10
1951	259.2	1941-51	49.9	—24.8	50
1956	279.6	1951-56	29.4	—9.1	37
		1921-56	137.6	—55.0	40
Females					
1921	190.5				
1931	199.6	1921-31	28.5	—18.9	68
1941	223.3	1931-41	29.2	—6.3	21
1951	256.5	1941-51	49.9	—16.8	34
1956	275.0	1951-56	30.3	—11.8	39
		1921-56	137.9	—53.7	39
Total					
1921	387.9				
1931	408.2	1921-31	56.5	—36.1	64
1941	457.4	1931-41	58.5	—10.2	17
1951	515.7	1941-51	99.8	—41.6	42
1956	554.6	1951-56	59.7	—20.9	35
		1921-56	275.5	—108.8	39
			Newfoundland		
Males					
1921	134.1				
1935	148.7	1921-35	22.5	—7,0	35
1945	164.6	1935-45	23.9	—8.0	33
1951	185.1	1945-51	25.8	—5.3	21
1956	213.9	1951-56	25.8	3.0	—12
		1921-56	98.0	—18.2	19
Females					
1921	128.9				
1935	140.9	1921-35	21.7	—9.7	48
1945	157.2	1935-45	23.9	—7.6	32
1951	176.3	1945-51	25.7	—6.6	26
1956	201.2	1951-56	26.0	—1.1	4
		1921-56	97.3	—25.0	26
Total					
1921	263.0				
1935	289.6	1921-35	44.2	—17.6	40
1945	321.8	1935-45	47.8	—15.6	33
1951	361.4	1945-51	51.5	—11.9	23
1956	415.1	1951-56	51.8	1.9	—4
		1921-56	195.3	—43.2	22

Sources: *Census of Canada, Vital Statistics,* and *Census of Newfoundland,* 1945, Report PN-2. See also *Canada Year Book 1959,* p. 163. For 1951-1956 see "Appendix on Statistical Method."

migration rates into the future. It appears that internal net migratory movements in Canada are surprisingly sensitive indicators of the geographical growing points of the economy. Thus, the influx of capital to Newfoundland since Confederation, together with a higher social security floor for persons dependent on seasonal and subsistence occupations may be expected to weaken the pressures which force people to leave home in search of a livelihood.

At the same time we wish to stress that net migration from the Maritimes has proceeded at very high rates, even in the nineteen fifties. This conclusion is reinforced by a different set of data based on a sample of unemployment insurance records.[3]

On the basis of these data, New Brunswick emerges with the highest provincial mobility rating in Canada. Of the insured persons found in New Brunswick in 1952, 22.1 per cent had left the province by 1956. Contrary to our own findings, Prince Edward Island shows a lower provincial mobility rating (17.5 per cent). This, we think, is a statistical illusion due to the fact that unemployment insurance data do not properly reflect movements off the farm. Nova Scotia shows a mobility rating of 13.1 per cent on this measure and Newfoundland shows 6.3 per cent. The Canada average is 7.1 per cent and the Ontario average is 4.1 per cent.

A finer breakdown which takes account of annual locational movement in the years 1952-56 between local unemployment insurance office areas again shows the New Brunswick insured employees as the most mobile in Canada.

Although we were unable to investigate occupational and industrial mobility on the basis of census data we note that whereas physical mobility out of the Atlantic provinces is very much higher than in other parts of Canada, average occupational mobility is the same (36 per cent).

The most interesting feature of occupational mobility patterns in the Atlantic region concerns the comparative rigidity at the top and the bottom of the occupational ladder. Persons in managerial and unskilled occupations are typically highly mobile in Canada. The Atlantic provinces depart from this pattern: managerial employees show a mobility rating of 39.1 per cent compared with a Canadian figure of 49.0 per cent, indicating a reluctance of people in managerial positions to relinquish them. Unskilled employees in the Atlantic provinces show a mobility rating of 38.2 per cent compared with a Canadian average of 52.4 per cent indicating a serious lack of upward occupational mobility at the bottom. Service, sales, and clerical workers on the other hand showed mobility rates higher than the Canadian average.

Birthplace Data

The census of Canada records the province or country of birth of the population found at the census date. We thus obtain a picture of the geographical location of persons born in the Atlantic region which reflects past migration patterns (see Table 3).

We found for instance, that only 76.9 per cent of people born in Prince Edward Island and living in Canada were living on the Island. 8.6 per cent were living in one of the other three Atlantic provinces, while 14.5 per cent were living else-

---

[3] D.B.S., Labour Division, *Movements within the Canadian Insured Population, 1952-56* (Ottawa, Feb., 1960).

TABLE 3

Canadian Residence of Persons Born in the Atlantic Provinces, 1951

Province of Residence, June, 1951

Province of birth	Nfld.	P.E.I.	N.S.	N.B.	Que.	Ont.	Man.	Sask.	Alta.	B.C.	Y.T.	N.W.T.	Can.
Newfoundland	353,478	269	14,173	1,461	5,260	19,124	412	292	773	2,321	24	36	397,263
Prince Edward Island	293	90,223	5,695	4,064	2,836	7,138	847	1,313	2,007	2,850	29	15	117,310
Nova Scotia	1,392	2,121	561,134	13,766	13,112	43,730	2,863	2,823	6,784	12,104	199	122	660,150
New Brunswick	512	1,188	14,461	459,755	27,305	32,027	1,690	1,649	3,612	7,631	93	61	549,984

Source: Census of Canada 1951, vol. I, Table 45.

where in Canada. We note that Nova Scotians and New Brunswickers who move out of their native province tend to move away from the Atlantic Region to other parts of Canada, while people migrating from Prince Edward Island and New-foundland were found in relatively larger numbers in the Atlantic region.

Table 3 shows the Province of residence of persons born in each of the four Atlantic provinces. The concentration of persons born in the Atlantic region is heaviest in Ontario with 102,019 people next is Quebec with 48,513, and third largest is British Columbia with 24,906 persons born in the Atlantic region. This table also shows the extent of past migration within the Atlantic area. We note, for instance, that there are more than twice as many Prince Edward Island-born persons living in Nova Scotia and New Brunswick than Nova Scotians and New Brunswickers living in Prince Edward Island. The exchange of population be-tween Nova Scotia and New Brunswick, however, is very even.

Birthplace data will also tell us the province or country of birth of the population in each of the four Atlantic Provinces in 1951 (see Table 4). The percentage of people residing in the four provinces who were born outside the Atlantic region are as follows: Newfoundland, 1.6 per cent; Prince Edward Island, 4.7 per cent;

TABLE 4

Birthplace of Population Residing in Canada and the Atlantic Provinces, 1951

	Province of residence				
Province of birth	New-foundland	Prince Edward Island	Nova Scotia	New Brunswick	Canada
Quebec	655	432	4,675	11,185	3,881,487
Ontario	809	844	8,321	3,616	3,645,074
Manitoba	102	156	1,155	520	699,587
Saskatchewan	150	270	1,583	630	817,404
Alberta	106	215	1,205	449	649,594
British Columbia	83	138	1,472	365	514,651
Yukon and Norhtwest Territories	7	2	30	11	16,654
United Kingdom and other British Commonwealth countries	2,084	1,015	14,396	9,820	933,049
United States	975	1,117	7,555	7,073	282,010
European and other countries	770	439	6,729	2,982	844,852
Atlantic Region	355,675	93,801	595,463	479,046	1,725,067
Total population	361,416	98,429	642,584	515,697	14,009,429

Source: *Census of Canada, 1951,* vol. I, Table 45.

Nova Scotia, 7.3 per cent; and New Brunswick, 7.1 per cent. The biggest element in Nova Scotia's population not born in the region consisted of British migrants, with Ontario-born persons in second place. For New Brunswick the largest element not born in the Atlantic region consisted of Quebec-born persons, with British immigrants in second place. The reader should bear in mind that these patterns are the result of historical migration patterns and that current trends might be quite different. While we do not know how many of the 994,562 Canadian-born persons living in the United States in 1950 migrated there from the Atlantic region, indirect evidence indicates that the old pattern of migration from the Maritimes to New England is receding.[4] The total number of Canadian-born persons found in the United States reached its peak in 1930 and showed successive declines in 1940 and 1950. It is estimated that in the decade 1931-41 there was an excess of Canadians returning from the United States over Canadians emigrating to the United States. During the war there was very little migration from Canada to the United States. While Massachusetts still leads all other states in the size of its Canadian-born population, there were 95,000 fewer Canadian-born persons in this State in 1950 than in 1930. All the New England states lost Canadian-born population, while a number of other states increased their Canadian-born population. The largest increases were in California, Washington, Oregon, Florida, and New Jersey.

The age composition of Canadian-born persons in the United States reflects the large migrations of Canadians thirty years ago. Almost 60 per cent of Canadian-born in the United States were over forty-five years of age and almost one-quarter were over sixty-five years of age. Since the end of the Second World War emigration to the United States has resumed at the rate of about 20,000 to 30,000 annually. The educational status of Canadian-born persons in the United States aged fourteen to twenty-four in 1950 reveals a specialized migration of highly trained persons. There is, however, no indication which region of Canada they came from. Twenty per cent of these young male Canadians had a college education compared with 11 per cent of the United States' population. These educational differentials were also found in Chicago, Los Angeles, Detroit, and New York, where educational attainment of the United States' population is much higher than the national average. Only in Boston was the educational attainment of Canadian-born persons lower (very much lower) than that of the average population. Correspondingly, the proportion of Canadians employed as craftsmen and foremen, operatives and kindred workers was 55.5 per cent in Boston, while 16.7 per cent of Canadians were employed in professional, technical, and managerial capacities. Similar figures for Chicago were 42.4 per cent and 33.8 per cent; Detroit 52.1 per cent and 23.6 per cent; Los Angeles 36.8 per cent and 35.6 per cent; New York 35.8 per cent and 36.3 per cent. This information does not, of course, prove that the Atlantic provinces lose a higher proportion of highly trained people to the United States than do other parts of Canada.

AGE COMPOSITION OF NET MIGRANTS

The most characteristic feature of net-migration is undoubtedly its age composition. As is to be expected, a very high proportion of net migrants are drawn from the 15-29 age group. Thus, 60 per cent of net out-migrants in Prince Edward Island came from the 15-29 age group. For New Brunswick the percentage was also 60 per cent.

To describe the age composition of net migrants we have defined an age-specific migration rate. This gets us around two difficulties: (a) The fact that we do not know the actual age of the intercensal migrant at the time of migration; (b) the

---

[4] D.B.S., *The Canadian-born in the United States,* Reference paper 71.

fact that the age distribution of the initial population will affect the age distribution of net migrants which it generates. The net migration rate for males aged 15-19 (1951-56) is thus defined as the number of people aged 15-19 in 1951 who had migrated by 1956 as a percentage of the base group. To take an example, the net migration rate for P.E.I. males, 15-19, in the period 1951-56 was 24.6 per cent. This tells us that nearly one quarter of males aged 15-19 in 1951 had left the province by 1956. These young men could have been aged anywhere between 15 and 24 at the time they left. The importance of choosing a measure independent of the age distribution of the initial population is illustrated by the fact that we find 639 male migrants aged 10-14 in 1951 and only 593 aged 20-24. Yet the migration rate shows that migration was proceding at a lower rate (13.4 per cent) among the younger group than among the older group (17.7 per cent).

Table 5 shows age-specific migration rates for males and females in the four Atlantic provinces. We note that age-specific migration rates for the 1941-51 period are higher than those for the 1951-56 period because they measure migration over a ten-year period. For comparison, the 1951-56 rates should be doubled. We consider these figures among the more interesting and important results of the study. For the sake of brevity we refrain from describing the results in the text and refer the reader to Table 5.

At this point we must warn the reader that it would not be correct to infer from the age distributions presented in our tables that there was in fact a net in-movement of older people to the Maritime Region. While we believe that these age distributions are reasonably accurate for the majority of age groups, they are not so with respect to the older groups. The reason is connected with the mortality rates which we used in estimating these distributions and are more fully explained in the "Appendix on Statistical Method."

TABLE 5

Age-Specific Net Migration Rates 1951-56

(By 5-year age groups, males and females, in percentages)

Age in 1951	Prince Edward Island		Nova Scotia		New Brunswick		Newfoundland	
	Male	Female	Male	Female	Male	Female	Male	Female
0-4	—4.0	—5.3	—0.1	0	—2.1	—1.6	1.0	2.6
5-9	—4.7	—4.9	—1.5	—1.4	—2.0	—2.6	—1.1	—0.1
10-14	—13.4	—9.5	—1.9	—3.5	—6.8	—2.4	—2.1	—0.3
15-19	—24.6	—26.0	—1.2	—7.2	—14.6	—14.2	3.3	—4.7
20-24	—17.7	—17.0	—0.8	—4.4	—2.2	—10.8	7.9	—4.0
25-29	—12.4	—11.5	—0.6	—2.6	—2.3	—5.7	4.6	—2.1
30-34	—5.6	—4.8	1.3	0.4	1.2	—0.8	9.0	4.1
35-39	—8.4	—4.6	—1.2	—2.8	—1.1	—3.1	2.3	—3.0
40-44	—1.0	—5.1	0.3	—2.5	—0.5	—2.0	3.3	—0.5
45-49	—1.9	—2.0	0.7	—2.2	—1.0	—3.1	1.7	—3.1
50-54	—3.5	—4.9	—1.2	—2.5	0.5	—1.8	1.8	—1.0
55-59	2.8	—1.1	0.6	—1.8	1.5	—2.9	—0.9	—3.5
60-64	8.3	3.6	5.6	3.6	5.6	4.2	5.8	4.0
65+	3.5	5.1	1.7	2.4	1.8	1.8	—2.0	—3.5

TABLE 5 (*continued*)

Age-Specific Net Migration Rates 1941-51

(By 5-year age groups males, and females, in percentages)

Age in 1941	Prince Edward Island		Nova Scotia		New Brunswick	
	Male	Female	Male	Female	Male	Female
0-4	0.2	—0.9	2.7	2.7	1.3	0.7
5-9	—12.7	—12.7	—7.0	—5.1	—11.3	—7.0
10-14	—28.7	—29.2	—16.3	—12.6	—26.0	—17.2
15-19	—28.4	—25.5	—16.6	—12.1	—24.2	—16.4
20-24	—25.8	—15.0	—14.6	—10.0	—18.1	—10.4
25-29	—14.6	—9.1	—7.8	—4.9	—7.7	—3.3
30-34	—11.3	—2.5	—3.5	—1.2	—2.5	—2.1
35-39	—7.8	—6.7	—6.2	—3.9	—4.9	—4.8
40-44	0.3	—2.8	—0.6	—0.2	—1.9	—2.7
45-49	—7.1	—6.1	—5.2	—3.9	—5.4	—4.7
50-54	—3.3	—7.8	—3.4	—5.3	—2.2	—4.3
55-59	4.5	1.0	5.8	2.6	3.6	1.2
60+	3.4	2.0	2.3	2.7	1.7	1.2

Source: Our estimates based on *Census of Canada, 1941*, vol. I, Table 22; *Census of Canada, 1956*, Bulletin 1-9; *Census of Canada, 1941*, vol. II, Table 21; *Census of Canada, 1951*, vol. I, Table 22. For seasonal rates used, see "Appendix on Statistical Method."

The higher physical mobility of young people is also revealed in the unemployment insurance data already referred to. These data show that 29.7 per cent of insured workers living in the Atlantic provinces in 1952 and aged under twenty had left the province by 1956. The same figure for insured workers aged 20-44 was 16.3 per cent and for workers aged 45-65 was 9.3 per cent. Provincial mobility for all age groups over the four year period was 15.5 per cent. Corresponding all-Canadian figures were 11.5 per cent, 8.0 per cent, and 4.0 per cent. The Canadian average was 7.2 per cent. Thus migration rates in the Atlantic provinces ran twice as high for the older groups and almost three times as high for the under-20 age group as in the rest of Canada.

Out-migration of young people from the Atlantic provinces has a very unfavourable effect on the age distribution of the region. The point was well made by Dr. S. A. Saunders in his Economic History of the Maritime Provinces:

For years, the movement of population has been out of rather than into the Maritime Provinces, with the result that the population is heavily weighted on the side of the very old and the very young leaving the number in the more productive age groups relatively lower than in other parts of the Dominion; and this, in itself, other things being equal, lowers the average income. From these circumstances flow two results that have a direct bearing upon government: first, there are relatively fewer taxpayers in proportion to population than in other regions of Canada; and second, such an age distribution of population tends to increase the burden of social services.[5]

Migration is however not the only factor determining the age distribution of the population. Changes in birth and death rates also affect the age distribution. There

[5] A Study Prepared for the Royal Commission on Dominion-Provincial Relations (Ottawa, 1939), pp. 103ff.

has been, in Canada, a disproportionate rise in those sections of the population which are either too young or too old to work. Thus, the proportion of persons of working age (15-64) in Canada as a whole declined from 65 per cent in 1941 to 60 per cent in 1956. The war and post-war "baby boom" together with a continuous fall in mortality rates are responsible for this change in age distribution.

The age distribution in the Maritime provinces is even less favourable. Out-migration from this region accentuates the adverse ratio of working to total population, whereas the inflow of migrants from other parts of Canada and from abroad alleviates the situation in those provinces which receive them. Thus while the proportion of people aged 15-64 to total population was 61.6 per cent in Ontario in 1956, it was 57.6 per cent in Nova Scotia, 55.2 per cent in New Brunswick, 54.7 per cent in Prince Edward Island and 53.3 per cent in Newfoundland.

Heavy net out-migration of people of working age provides one explanation for this situation. In the case of P.E.I., for example, this factor alone explains the distorted age distribution. Differentially higher crude birth rates, however, such as are found in Newfoundland and New Brunswick accentuate the problem. In the case of Newfoundland, half the population (53.3 per cent) has to support the other half (46.7 per cent) whereas in Ontario these figures are 61.6 per cent and 38.4 per cent. The reason for the high proportion of children in Newfoundland is found in its high crude birth rate, of 33.8 per 1,000 in 1958 (the highest in any province of Canada).[6] In New Brunswick, both out-migration and a high birth rate act to distort the age distribution. Table 6 illustrates the age distribution

TABLE 6

Percentage Distribution of the Population by Specified Age Groups for Ontario and the Atlantic Provinces

	1941	1951	1956		1941	1951	1956
	Newfoundland				New Brunswick		
Age group				Age group			
0-14	—	39.1	40.7	0-14	31.0	35.7	37.0
15-64	—	54.4	53.3	15-64	61.0	56.7	55.2
65	—	6.5	6.0	65	7.1	7.6	7.8
Total	—	100.0	100.0	Total	100.0	100.0	100.0
	Prince Edward Island				Ontario		
0-14	30.2	33.4	34.9	0-14	24.4	27.0	30.0
15-64	60.4	56.7	54.7	15-64	61.7	64.3	61.6
65	9.4	9.9	10.4	65	8.0	8.7	8.4
Total	100.0	100.0	100.0	Total	100.0	100.0	100.0
	Nova Scotia				Canada		
0-14	29.2	32.7	33.9	0-14	27.8	30.4	32.4
15-64	62.6	58.8	57.6	15-64	65.5	61.8	59.9
65	8.2	8.5	8.5	65	6.7	7.8	7.7
Total	100.0	100.0	100.0	Total	100.0	100.0	100.0

Source: *Census of Canada, 1956*, Bulletin 3-3.

[6] Out-migration of young adults would in fact reduce the crude birth rate in Newfoundland. Age specific fertility rates in Newfoundland were *lower* than in P.E.I. Yet P.E.I. had a crude birth rate of only 25.8 in 1958. The reason is to be found in the heavy out-migration of females of child bearing age.

of the population of the Atlantic provinces in comparison with that of the total Canadian population.

Another, and perhaps more dramatic illustration of the same set of facts is provided by Table 7. Here we note that there has been an *absolute decline* in the numbers of young persons aged 15-29 in each of the three Maritime provinces

TABLE 7

Percentage Increase in Population by Age Groups, Canada and Provinces, 1941-56

Age group		Canada	Ontario	Prince Edward Island	Nova Scotia	New Brunswick
Total		36.1	42.7	4.5	20.2	21.3
	0-14	58.1	75.1	20.8	39.5	40.8
	15-19	0.6	2.3	—10.8	—0.2	—2.9
	20-24	6.5	12.7	—25.6	—10.7	—13.9
	25-29	21.1	32.2	—27.8	—10.2	—6.8
	30-44	44.0	47.9	8.1	30.6	28.8
	45-49	26.7	30.2	2.8	18.5	18.3
	60-64	26.3	29.8	11.5	16.4	22.3
	65	58.8	50.8	15.6	25.3	33.2
0-14 & 65 and over		58.2	69.1	19.5	36.4	39.4
15-64		24.5	30.1	—5.4	10.5	9.7
Ratio of percentage gain in non-working age groups over percentage in 15-64 group		2.4	2.3	loss	3.5	4.1
Same ratio for period, 1951-56		2.0	2.1	loss	3.5	2.5

Source: *Census of Canada, 1956*, Bulletin 3-3.

in the period 1941-56. Among persons aged 20-24 there was a decline of 26 per cent in P.E.I.; 14 per cent in New Brunswick; 11 per cent in Nova Scotia although this same group *increased* by 6.5 per cent in Canada and by 13 per cent in the Province of Ontario. Over the period 1941-56 people in the non-working age groups increased 3.5 times as fast as those in the working age groups in Nova Scotia and 4.1 times as fast in New Brunswick, while in Canada the working population increased only 2.4 times as fast as the non-working population.

This unfavourable age distributions acts to depress income per person in the Atlantic provinces. The impending change in the age distribution as a result of the current and future enlarged rate of entry of school leavers into the labour market will not, however raise per person incomes in the Atlantic region unless there is a corresponding increase in the rate at which new employment opportunities are created. Pending such an increase, there is bound to be more unemployment and more underemployment in the future. This promises to present an especially serious problem in Newfoundland where 40 per cent of the population was under 15 in 1956, as against 30 per cent in Ontario. Lower educational levels and the restricted range of employment opportunities associated with the rudimentary state of the economy of Newfoundland threaten to create a very serious problem there.

CULTURAL DIFFERENTIALS IN NET MIGRATION RATES

By applying survival rates to census data of population classified according to origin, mother tongue and religious denomination, it is possible to investigate differences in net migration rates between ethnic and religious groups.

This exercise was performed on data for the 1941-51 period because the 1956 census did not provide the required breakdown. The procedure rests on two assumptions: (a) origin, mother tongue and religious denomination are fixed characteristics in the sense that it is assumed that an individual gave the same answer to questions concerning these attributes in the 1951 census as he gave in the 1941 census—an assumption dubious in the case of religious denomination; (b) there are no differences in mortality rates between various ethnic and religious groups of immigrants.[7]

The results are presented in Tables 8, 9, and 10. We found that the population of French origin had a higher net out-migration rate than the population of British origin. We found this to be so for both men and women and in all three Maritime provinces. There was no exception to this result. The difference between French and British net-migration rates was most striking in Prince Edward Island and was smallest in Nova Scotia. We considered the possibility that the apparent differences in net migration rates are really differences in mortality rates as between persons of British and French origin. We rejected this explanation because such differences in mortality rates if they exist, would show up at older ages.

But migration typically takes place at young age. We have migration rates for individual age-groups where mortality is anyway very low. Thus in Prince Edward Island, 28.9 per cent of males of British origin aged 10-14 in 1941 had migrated by 1951, whereas 35.8 per cent of males of French origin aged 10-14 in 1941 had left the province by 1951. Among the male age group 15-19 in 1941 the migration rate for migrants of British origin was 26.8 per cent whereas it was 36.8 per cent for males of French origin. As we said, mortality is so low at these ages that it is impossible that these differences could be caused by our use of the same mortality factor for both groups (see Table 11).

The data we have used in this analysis do not permit any inference as to the reasons for the differential net migration rates we have yourd. We cannot say that Canadians of French origin have a higher migration rate because they are French Canadians. It is possible that they have a higher migration rate because they are located in areas which for economic reasons produce heavier out-migration. This might be a plausible explanation if the differential were found only in New Brunswick. It is however found also in the other two Maritime provinces. Another explanation might be the higher birth rates among Canadians of French origin. More to the point, we must remined the reader of the limitations of the concept of net-migration. Thus it is possible that persons of British and French origin leave the area at the same rate but there is a larger counter balancing inflow of migrants of British origin. This indeed is very likely. One thing, however, we can state with certainty: the reason for the relative increase of the French-Canadian population in the Maritimes is *not* to be found in a lower propensity to migrate out of the area, but in a higher birth rate.

---

[7] For further details on procedure, see "Appendix on Statistical Method."

TABLE 8

Migration Rates by Origin of Net Migrants, 1941-1951

	Males			Females		
	Number	Per cent*	Net migration rate	Number	Per cent*	Net migration rate
Prince Edward Island						
1941 Population						
British	40,730	82.7	84.2	—	37,984	82.9	84.1	—
French	7,619	15.5	15.8	—	7,180	15.7	15.9	—
Other	879	1.8	—	—	655	1.4	—	—
Total	49,228	100.0	—	—	45,819	100.0	—	—
Migrants 1941-1951						
British	—4,583	—	78.7	—11.3	—3,760	—	78.8	—9.9
French	—1,307	—	21.3	—17.2	—1,013	—	21.2	—14.1
Other	187	—	—	21.3	113	—	—	20.3
Total	—5,703	—	—	—11.5	—4,640	—	—	10.1
Nova Scotia						
1941 Population						
British	226,775	76.6	86.9	—	218,403	77.4	87.2	—
French	34,069	11.5	13.1	—	32,191	11.4	12.8	—
Other	35,200	11.9	—	—	31,324	11.2	—	—
Total	296,044	100.0	—	—	281,918	100.0	—	—
Migrants 1941-1951						
British	—18,216	—	86.4	—8.0	—13,543	—	86.7	—6.2
French	—2,868	—	13.6	—8.4	—2,071	—	13.3	—6.4
Other	2,371	—	—	6.7	2,728	—	—	8.7
Total	—18,713	—	—	—6.3	—12,886	—	—	—4.5
New Brunswick						
1941 Population						
British	141,360	60.4	62.8	—	135,398	60.6	62.7	—
French	83,503	35.7	37.1	—	80,431	36.0	37.2	—
Other	9,234	3.9	—	—	7,475	3.4	—	—
Total	234,097	100.0	—	—	223,304	100.0	—	—
Migrants 1941-1951						
British	—13,886	—	61.4	—9.8	8,841	—	55.8	—6.5
French	—8,747	—	38.6	—10.5	7,008	—	44.2	—8.7
Other	1,012	—	—	11.0	1,683	—	—	22.5
Total	—21,611	—	—	—9.2	—14,166	—	—	—6.3

* On the left side we show percentage distribution of the total 1941 population. On the right side we show percentage distribution of the sum of British and French components only.

Source: *Census of Canada, 1941*, vol. III, p. 140; *Census of Canada, 1951*, vol. II, Table 5. Using Maritime male and female survival rates, 1941, to estimate migration. For method see "Appendix."

TABLE 9

Migration Rates by Mother Tongue of Net Migrants, 1941-1951

	Males			Females	
	Number	Per cent	Net migration rate	Number	Net migration rate
Prince Edward Island					
1941 Population					
English	43,124	87.6	—	40,118	—
French	5,478	11.1	—	5,200	—
Other	626	1.3	—	—	—
Total	49,228	100.0	—	45,318*	—
Migrants 1941-1951					
English	—4,111	72.1	—9.5	—3,221	—8.0
French	—1,492	26.1	—27.2	—1,288	—24.8
Other	—100	1.8	—16.0	—	—
Total	—5,703	100.0	—11.6	—4,509*	—10.1
Nova Scotia					
1941 Population					
English	262,097	88.5	—	251,946	—
French	21,237	7.2	—	20,113	—
Other	12,710	4.3	—	—	—
Total	296,044	100.0	—	272,059*	—
Migrants 1941-1951					
English	—13,489	72.1	—5.1	—8,092	—3.2
French	—2,634	14.1	—12.4	—2,632	—13.1
Other	—2,590	1.4	—20.4	—	—
Total	—18,713	100.0	—6.3	—10,724*	—4.5
New Brunswick					
1941 Population					
English	149,723	64.0	—	143,616	—
French	80,639	34.0	—	63,940	—
Other	3,735	1.6	—	—	—
Total	234,097	100.0	—	207,556*	—
Migrants 1941-1951					
English	—11,020	51.0	—7.4	—5,694	—4.0
French	—9,627	44.5	—11.9	—8,155	—10.6
Other	—974	4.5	—26.1	—	—
Total	—21,621	100.0	—9.2	—13,849*	—6.3

* English and French speaking population and migrants only.

Source: *Census of Canada, 1941*, vol. III, Table 42. *Census of Canada, 1951*, vol. II, Table 22. Using Maritime male and female survival rates, 1941, to estimate migration. For method see "Appendix."

TABLE 10

Migration Rates by Religious Denomination of Net Migrants, 1941-1951

	Males			Females		
	Number	Per cent	Net migration rate	Number	Per cent	Net migration rate
Prince Edward Island						
1941 Population						
Roman Catholic	22,275	45.2	—	20,468	44.7	—
Protestant*	26,953	54.8	—	25,351	55.3	—
United Church	12,314	—	25.0	—	—	—
Presbyterian	7,596	—	15.4	—	—	—
Anglican	2,914	—	5.9	—	—	—
Baptist	2,822	—	5.7	—	—	—
Total	49,228	100.0	—	45,819	100.0	—
Migrants 1941-1951						
Roman Catholic	—3,374	59.2	—15.1	—2,727	58.7	—13.3
Protestant*	—2,329	40.8	—8.6	—1,913	41.3	—7.6
Total	—5,703	100.0	—11.6	—4,640	100.0	—10.1
Nova Scotia						
1941 Population						
Roman Catholic	96,839	32.7	—	92,105	32.7	—
Protestant*	199,205	67.3	—	189,813	67.3	—
United Church	62,631	—	21.2	—	—	—
Presbyterian	24,887	—	8.4	—	—	—
Anglican	52,810	—	17.8	—	—	—
Baptist	45,631	—	15.4	—	—	—
Total	296,044	100.0	—	281,918	100.0	—
Migrants 1941-1951						
Roman Catholic	—7,675	40.0	—7.9	—5,201	40.0	—5.6
Protestant*	—11,037	60.0	—5.5	—7,685	60.0	—4.0
Total	—18,713	100.0	—6.2	—12,886	100.0	—4.5
New Brunswick						
1941 Population						
Roman Catholic	112,864	48.2	—	107,590	48.1	—
Protestant*	121,233	51.8	—	115,714	51.9	—
United Church	31,830	—	13.6	—	—	—
Presbyterian	8,171	—	3.5	—	—	—
Anglican	28,150	—	12.0	—	—	—
Baptist	45,627	—	19.5	—	—	—
Total	234,097	100.0	—	223,304	100.0	—
Migrants 1941-1951						
Roman Catholic	—11,902	55.0	—10.6	—8,976	63.0	—8.3
Protestant*	—9,719	45.0	—8.0	—5,279	37.0	—4.6
Total	—21,621	100.0	—9.2	—14,255	100.0	—6.4

* Includes Orthodox and non-Christian religious denominations. These are very small in the Maritime Provinces.

Souce: *Census of Canada, 1941*, vol. III, Table 14, *Census of Canada, 1951*, vol. II, Table 8. Using Maritime male and female survival Rates, 1941, to estimate migration. For method see "Appendix."

TABLE 11

Net Out-Migration Rates for Selected Male Age Groups, and Cultural Characteristics, 1941-1951

(in percentages)

Age group		Prince Edward Island		Nova Scotia		New Brunswick	
				Origin			
	At time of migration	British	French	British	French	British	French
15-19	5-19	12.1	20.6	8.4	10.0	12.0	11.7
20-24	10-24	28.9	35.8	17.7	19.9	25.4	29.4
25-34	15-34	26.9	36.8	16.8	21.6	21.2	23.8
				Mother Tongue			
		English	French	English	French	English	French
15-19	5-19	9.1	39.9	5.7	14.2	9.0	13.9
20-24	10-24	26.3	49.2	15.8	17.7	22.7	31.2
25-34	15-34	25.3	45.6	14.0	26.7	18.5	25.2
				Religious Denomination			
		Roman Catholic	Protestant	Roman Catholic	Protestant	Roman Catholic	Protestant
15-19	5-19	16.5	8.9	8.6	6.1	11.7	10.7
20-24	10-24	34.5	23.0	15.2	15.3	28.3	23.3
25-34	15-34	33.2	21.8	17.4	14.7	23.6	18.9

Source: From our own computations based on Tables 8, 9, and 10.

An analysis of migrants by mother tongue bears out the findings of the analysis of origin. The differentials are much wider and once again general. Some reasons for the widening of differentials as between the origin and the mother tongue analyses suggest themselves. Many people of ethnic origin other than British have English for mother tongue. The analysis of migrants by origin showed net in-migration of persons of origin other than British or French (see Table 8). A large number of these persons might be English-speaking. This would depress the rate of net out-migration of English-speaking persons. This rate is indeed everywhere lower than that for the British-origin category.

Now we note that the difference between the French migration rate based on the origin count and that based on the language count is not very large in New Brunswick (—10.5 per cent and —11.9 per cent). In New Brunswick the vast majority of people of French origin are also French-speaking. The difference is much larger in Nova Scotia (—8.4 per cent and —12.4 per cent) and Prince Edward Island (—17.2 per cent and —27.2 per cent). But in Nova Scotia and to a lesser extent in Prince Edward Island a very considerable number of persons of French descent are English-speaking (see Tables 8 and 9). It thus appears that the French speaking people of French origin have a higher net migration rate than those who are English-speaking. This reinforces the argument we made above.

Table 10 presents migration rates by religious belief of migrants. We note that Roman Catholics have a higher net migration rate than Protestants and that the differentials are again found for both men and women and in all provinces and are again greatest in Prince Edward Island. If we compare the religious with the origin analysis for New Brunswick we could infer that it is typically Protestants of British origin who have the lower net migration rate. The migration rate for Protestants is lower than that for people of British origin.

In summary on a net basis at least, Canadians of French origin in the Maritimes are *more, rather than less,* mobile than Canadians of British origin. Further, there is some evidence that, among people of British origin, those who are of Protestant denomination are more tenacious or successful in remaining, perhaps because of educational or social advantages of this group.

MIGRATION AND EDUCATIONAL ATTAINMENT

It is a commonly held belief that out-migration is most pronounced among persons with above-average educational attainment. We found that the census data would not stand up to a proper statistical test of this hypothesis. We hope that the additional information which, we understand, will be made available in connection with the 1961 census, will yield a definite answer to this rather important question. The hypothesis is probably true in the sense that, all other factors being equal, educational attainment is correlated with out-migration. Such is the common observation which we share. As we shall try to show later, however there are reasons for believing that economic and social pressures generate a differentially higher rate of out-migration among groups of people whose average educational attainment is lower rather than higher. We shall return to this discussion later.

First we will describe some data on out-migration of university graduates from this region. On the basis of information supplied by the Alumni Offices of three major universities in the region we can observe the present place of residence of nine generations of university graduates. The results we present refer to graduates whose original home was in the Atlantic region. We thus excluded all graduates who came to these universities from other parts of Canada or from abroad. It is understandable that no alumni office can have complete records of the present residence of past graduates. The percentages we record are thus based on records of those graduates whose original home was in the Atlantic region and whose present place of residence is known and recorded.[8] They refer to the years 1950 to 1958 inclusive.

We found that 50 per cent of graduates of the University of New Brunswick are presently living in the Atlantic region while 59 per cent of graduates of Dalhousie University, and 64 per cent of graduates of St. Francis Xavier University, are residing in the region. In all three cases, the percentage was higher for Arts graduates and lower for Science and Engineering graduates. At the University of New Brunswick for example, 60 per cent of Arts graduates were living in the region, whereas the percentages for some of the other faculties were: Civil En-

---

[8] Data for 1959 were not used because they were not considered reliable as many 1959 graduates have not yet settled permanently. The same applies even more forcefully to 1960 graduates. Our estimates of the percentage of graduates who have left the Province is likely to be conservative because it is probable that the people who have left among those whose address was *not known* was higher.

gineering, 52 per cent; Forestry, 48 per cent; Science, 40 per cent; Mechanical Engineering, 39 per cent and Electrical Engineering, 38 per cent. At St. Francis Xavier University 67 per cent of Arts graduates were residing in the region while only 57 per cent of Science graduates were presently in the region. Those who were not living in the region were mainly found in the rest of Canada, with a small number living abroad. Thus in the case of the University of New Brunswick, 42 per cent of graduates were living elsewhere in Canada and 8 per cent were living abroad. No trend can be inferred from the annual data covering the nine graduation years. We say this since one would expect the graduates of 1950 to be more scattered than those of later years—because more time has elapsed. Further, the higher percentage of people resident in the region shown for 1958 is not considered reliable as some of these graduates are probably still receiving mail at their original home address.

Thus we note that a very considerable proportion of Maritime graduates have left the Maritimes. Without similar data on the present whereabouts of graduates of other Canadian universities it is not however possible to say whether local Maritime graduates tend to scatter more than, say, Ontario men and women graduating from a major Ontario university.

It was when we tried to deal with the vast majority of out-migrants who did not receive a university education that we ran into severe statistical difficulties arising from a lack of data. Obviously migration is very strongly affected by age. In so far as young people have, on average, more education than their elders, the educational attainment must be, on average, higher than that of the remaining population. Thus for example in New Brunswick, in 1951, 54.7 per cent of persons aged 5-24 were attending school, whereas in 1941 only 49.2 per cent of this age group were in school. In 1951, 44.6 per cent of persons aged 20-24 and not attending school had undergone nine or more years of schooling. A similar percentage for persons aged 25-34 in 1951 was 42.6 per cent; for persons aged 35-44, 38.6 per cent; and persons aged 45-64, 34.2 per cent. This is a measure of the improvement in educational facilities over a long period of time.[9]

The question we would really have liked to have been able to answer is whether, in any given age-group, the migration rate for those with more schooling is higher —or lower—than for those with less schooling. In other words, whether education, taken by itself and apart from the effect attributable to the age distribution of migrants, does or does not affect out-migration. Our rather elaborate efforts to estimate this by using the survival method on census data by years of schooling failed because we were unable to find information to make a reliable estimate of what the distribution by years of schooling for persons attending school in 1941 would have been by 1951, if none of these persons had migrated.

We believe it probable that, within any homogeneous group of young people, those with more schooling are more likely to migrate. As we said, no test can be devised for this hypothesis on the basis of existing data. Such a partial correlation, if it exists, could however, easily be counteracted by the very strong rural origin of net migration which we shall describe in another section of this paper. In so far as only 31.6 per cent of rural young people in New Brunswick

---

[9] Years of schooling as used in the census refer to the number of years spent at school rather than grade achieved. Credit is awarded for night courses, special day courses, etc.

had more than eight years of schooling in 1951 (see Table 12) compared with 60.6 per cent of urban young people, it is very probable that the total group of provincial net out-migrants of a given age will have less schooling than young people of the same age who remain within the province. Similar and related large

TABLE 12

Educational Attainment of Persons of Selected Age Groups, Rural and Urban, New Brunswick, 1951

	Number of years schooling					
	None	1-4	5-8	9-12	13	Total
Age 20-24						
Rural	538	2,195	10,574	5,720	424	19,451
Urban	110	476	5,598	8,553	979	15,716
Percentage rural	2.7	11.3	54.4	29.4	2.2	100.0
Percentage urban	0.7	3.0	35.6	54.4	6.2	100.0
Age 25-34						
Rural	436	4,634	21,354	10,019	996	38,139
Urban	274	1,326	12,439	16,692	2,810	33,541
Percentage rural	3.0	12.1	56.0	26.3	2.6	100.0
Percentage urban	0.8	4.0	37.1	49.7	8.4	100.0

Source: *Census of Canada, 1951*, vol. II, Table 28.

differentials between educational attainment of persons of British and French origin (see Table 13) taken in conjunction with the already established differentially higher rates of net out-migration of persons of French origin point to the same conclusion: that average educational attainment among net out-migrants may be lower than that prevailing among young people of the same age remaining in the province.

TABLE 13

Educational Attainment of Persons Five and Over and Not Attending School, by Origin, New Brunswick, 1951

	Number of years schooling					
	None	1-4	5-8	9-12	13	Total
British origin	11,515	12,040	86,847	80,129	10,271	200,802
French origin	18,778	24,563	56,244	18,267	2,368	120,220
Percentage British	15.7	6.0	43.3	39.9	5.1	100.0
Percentage French	15.6	20.4	46.8	15.2	2.0	100.0

There is no cross-classification of education attainment by origin for separate groups. The differential between *young* people of British and French origin is certainly smaller than that for the whole population in this table.

Source: *Census of Canada, 1951*, vol. II, Table 52.

This conclusion does not constitute a contradiction of the observed fact that the migration rate of out-migration among professional people is very high. That this is so we noticed from our analysis of the present residence of university graduates. It can also be seen in the fact that 5.8 per cent of insured persons classed as "professional" moved out of the Atlantic region in the single year 1954-55.[10] No other region of Canada recorded such a high rate of loss of professional workers. The rate of loss among clerical workers of 3.6 per cent in the same year was also unequalled by any other region. However, the number involved in these (highly educated) occupations is much smaller than the number of (poorly educated) labourers recorded as having moved in the same year. We thus reiterate our suggested conclusion that it is doubtful whether the educational distribution of net migrants as a whole differs from that of the remaining population, once we have allowed for age-distribution.

ECONOMIC AND SOCIAL ORIGIN OF MIGRATION

In this concluding section of the paper we shall summarize some of our findings concerning the relationship between net migration rates and the economic and social structure of the region. Table 14 summarizes the main relations while detail by counties is given in Tables 15 and 16. The following summary conclusions may be drawn from these data.

TABLE 14

Summary of Key Socio-economic Indicators Related to Net Migration Rates

	P.E.I.	N.B.	N.S.
1. Net migration, 1941-51	—12,400	—41,653	—38,738
2. Net migration, 1951-56	— 8,107	—20,711	—10,687
3. Net migration, 1941-51 as percentage of 1941 population	—13.0%	—9.1%	—6.7%
4. Net migration, 1951-56 as percentage of 1951 population	— 8.2%	—4.0%	—1.7%
5. Estimated net migration from rural farm population (1951-56)	— 7,000	—30,000	—20,000
6. Above as percentage of rural farm population, 1956	16 %	24 %	21 %
7. Percentage of population classified rural, 1951	69.3%	54.2%	42.6%
8. Percentage of males in primary industry, 1951	51.5%	35.7%	30.2%*
9. Percentage of males earning less than $1500, 1951	57.9%	44.2%	38.0%
10. Estimated earned income per head (1951-55)† (Canada = 100)	49.2	63.5	70.1

* This figure includes coal miners. If they were excluded, the percentage would be 22.8 per cent.

† Howland: *Some Regional Aspects on Canada's Economic Development*, p. 45.

1. Internal net migration is primarily an aspect of the process of urbanization and the shift from agriculture and fishing to industrial and tertiary occupations. Whereas rural farm population in Canada declined by 5 per cent in the 1951-56 period, the rural farm population of Nova Scotia declined by 13.4 per cent, New

[10] R. D. Howland, *Some Regional Aspects of Canada's Economic Development* (Ottawa, 1958), p. 198.

TABLE 15

Percentage Increase of Population by Census Periods 1901-1956
Net Migration and Net Migration Rates
During the Periods 1941-1951 and 1951-1956

	Percentage population Increase between censuses				Net migration 1941-51	Net migration rate 1941-1951	Net migration 1951-56	Net migration rate 1951-56
	1901-11	1911-21	1921-31	1931-41				
			New Brunswick					
Heavy out-migration								
Victoria	30.8	10.9	16.5	11.8	-2,324	-13.9*	-2,149	-11.6*
L Kent	1.7	-1.9	-1.8	10.0	-5,280	-20.5*	-2,667	-10.0*
L Queens	-2.5	7.2	-3.9	13.9	-1,634	-12.8*	-1,270	-9.6*
L Charlotte	-5.1	-1.4	-0.5	6.5	-332	-1.5*	-2,313	-9.2*
H Madawaska	35.5	20.7	21.8	14.9	-3,192	-11.3*	-2,390	-7.0*
L Carleton	-0.8	-1.6	-1.4	4.4	-2,775	-12.8*	-1,397	-6.3*
Restigouche	48.2	45.6	30.7	10.8	-5,982	-18.1*	-2,051	-5.7*
H Gloucester	16.7	18.4	8.3	19.1	-8,270	-16.6*	-2,524	-4.4*
Northumberland	9.3	8.9	0.4	12.8	-4,225	-11.0*	-1,549	-3.¹
Mixed migration								
Westmorland	6.1	19.6	7.7	12.1	+177	+0.3	-3,618	-4.5*
Saint John	3.5	12.9	1.9	11.7	-5,299	-7.7	+191	+0.3
York	-0.2	2.2	0.6	12.3	-429	-1.2	+429	+1.0
Kings	-4.9	-0.9	-2.9	8.9	-1,458	-6.8	+285	+1.3
Sunbury	8.6	-0.9	13.6	18.5	-731	-8.8	+196	+2.1
In-migration								
Albert	-11.3	-11.2	-10.8	9.7	101	1.2	+116	+1.2
New Brunswick	6.3	10.2	5.2	12.0	-41,653	-9.1	-20,711	-4.0

TABLE 15 (continued)

	Percentage population Increase between censuses				Net migration 1941-51	Net migration rate 1941-1951	Net migration 1951-56	Net migration rate 1951-56
	1901-11	1911-21	1921-31	1931-41				
Nova Scotia								
Heavy out-migration								
L Guysborough	−6.9	−9.0	−0.5	0.1	−3,201	−20.7*	−1,476	−10.4*
L Yarmouth	−1.5	−3.6	−6.4	7.0	−3,372	−15.0*	−2,073	−9.1*
L Victoria	−6.3	−10.2	−11.0	1.3	−703	−8.8*	−666	−8.1*
L Inverness	5.0	−6.9	−11.6	−2.3	−4,497	−21.9*	−1,445	−7.8*
L Digby	−0.8	−2.8	−6.7	6.1	−2,223	−11.4*	−1,556	−7.8*
Cape Breton	49.0	17.7	7.2	19.7	−14,087	−12.7*	−8,963	−7.4*
L Cumberland	12.0	1.6	−11.7	8.6	−5,613	−14.2*	−2,785	−7.0*
L Richmond	−1.8	−6.1	−11.0	−2.2	−1,622	−14.9*	−742	−6.9*
Pictou	7.2	13.9	−4.5	4.5	−3,458	−8.5*	−2,765	−6.3*
L Shelburne	−0.7	−4.4	−7.5	6.1	−1,138	−8.6*	−915	−6.3*
Queens	−1.2	−1.6	6.7	13.3	−1,443	−12.0*	−661	−5.3*
Antigonish	−12.2	−3.2	−13.0	4.7	−326	−3.1	−313	−2.6*
Hants	−1.8	0.2	−1.8	13.6	−2,339	−10.6*	−591	−2.5*
L Lunenburg	2.7	1.4	−6.1	4.0	−3,034	−9.2*	−769	−2.3*
Mixed migration								
L Annapolis	−1.4	−2.3	−10.2	8.7	+1,700	+9.6	−1,569	−7.2*
Colchester	−5.0	6.5	−0.6	20.3	−3,622	−12.0*	+463	+1.4
H Kings	−0.7	8.9	2.7	18.7	−900	−3.0	+1,079	+3.3
In-migration								
H Halifax	7.5	21.0	3.1	22.4	11,140	9.0	+15,060	+9.3
Nova Scotia	7.1	6.4	−2.1	12.7	−38,738	−6.7	−10,687	−1.7

TABLE 15 (continued)

Prince Edward Island

	Percentage population Increase between censuses				Net migration 1941-51	Net migra-tion rate 1941-1951	Net migration 1951-56	Net migra-tion rate 1951-56
	1901-11	1911-21	1921-31	1931-41				
H Prince	−7.4	−3.8	−0.1	9.5	−3,958	−11.5	−4,051	−10.7*
L Kings	−8.4	−9.7	−6.3	1.4	−4,192	−21.6*	−1,480	−8.2*
H Queens	−11.2	−4.3	2.0	10.0	−4,250	−10.3	−2,576	−6.0
Prince Edward Island	−9.2	−5.5	−0.7	8.0	−12,400	−13.0	−8,107	−8.2

Newfoundland

	Percentage population				Actual increase 1945-51	Natural increase 1945-51	Net migration 1945-51	Net migration 1951-56	Net migra-tion rate 1951-56
	1901-11	1911-21	1921-35	1935-45					
Division 3		Not available			−0.3			−1,442	−7.0*
Division 7								−1,749	−4.9*
Division 2					1.9			−834	−3.7*
H Division 9					15.0			−485	−2.8*
H Division 4					22.3			−428	−2.7*
Division 8					11.4			+64	+0.2*
Division 1					10.5			+1,303	+0.9
H Division 6					40.5			+1,676	+6.0
H Division 5					35.6			+1,911	+6.8
H Division 10					43.0			+1,717	+21.8
Newfoundland	9.8	8.4	10.1	11.1	12.3	17.0	−4.7	1,733	+0.5

* Out-migration above provincial average.

L and H indicate trend of low and high growth rates since 1901.

Source: *Census of Canada, 1956*, Bulletin 3-1, Tables 1 and 2. *Census of Canada, 1951*, vol. X, Table 3. For Newfoundland see *Census of Newfoundland, 1901-45*; *Census of Canada, 1951*, vol. I, Table 6 for 1951.

KARI LEVITT

TABLE 16

Some Socio-Economic Indicators, by Counties, 1951

County	Population 1956	Percentage rural in 1956	Percentage male wage earners earning $1499 or less	Percentage of farms classified as commercial 1956	Percentage males in primary industry	Percentage growth of population by natural increase 1951-56
Prince Edward Island						
Kings	17,853	85.4	72.6	65.6	68.9	7.7
Prince	38,007	78.0	59.2	74.1	52.0	11.4
Queens	43,425	55.1	52.0	78.5	46.5	7.6
Total	99,285	69.3	57.9	73.8	52.9	9.1
Nova Scotia						
Annapolis	21,682	87.0	37.3	43.0	29.2	6.9
Antigonish	13.076	72.5	54.6	48.0	48.2	11.8
Cape Breton	125,478	11.7	18.8	30.4	39.0	11.7
Colchester	34,640	54.1	47.3	52.7	28.0	8.4
Cumberland	39,598	42.9	46.0	49.1	43.8	6.9
Digby	19,869	89.2	72.9	15.5	36.6	7.2
Guysborough	13,802	82.0	70.6	23.2	46.0	7.3
Halifax	197,943	17.0	27.0	37.0	5.2	12.7
Hants	24,889	80.1	56.9	55.1	44.2	9.1
Inverness	18,235	76.9	64.2	36.7	60.5	7.0
Kings	37,816	73.6	57.1	62.4	38.0	10.7
Lunenburg	34,207	72.3	59.0	18.9	37.6	5.2
Pictou	44,566	38.3	37.7	41.3	32.8	7.6
Queens	12,774	46.5	47.1	26.3	24.2	7.1
Richmond	10,961	100.0	63.3	22.9	46.5	8.5
Shelburne	14,604	75.7	70.3	30.6	43.5	7.8
Victoria	8,185	100.0	61.4	21.5	50.6	7.7
Yarmouth	22,392	63.8	60.0	20.4	32.6	7.3
Total	694,717	42.6	38.0	38.5	30.2	9.8
New Brunswick						
Albert	10,943	85.3	43.5	52.4	29.6	9.2
Carleton	23,073	76.9	59.9	72.3	53.2	9.9
Charlotte	24,497	59.3	46.1	39.2	38.4	6.7
Gloucester	64,119	81.8	62.0	14.9	53.0	15.9
Kent	27,492	87.9	75.4	42.8	66.7	12.7
Kings	24,267	60.4	51.9	64.8	43.7	6.7
Madawaska	36,988	63.3	45.1	52.9	46.3	14.7
Northumberland	47,223	71.8	57.3	15.3	43.8	13.4
Queens	12,838	90.7	49.8	50.6	59.5	6.8
Restigouche	39,720	54.5	42.4	27.1	34.6	15.4
Saint John	81,392	2.0	29.7	23.6	3.5	9.0
Sunbury	10.547	85.3	58.0	34.6	69.0	11.0
Vicotria	19,020	73.3	57.8	69.4	50.0	14.2
Westmorland	85,414	37.4	32.1	42.4	18.3	11.3
York	47,083	43.0	39.8	48.2	32.3	9.7
Total	554,616	54.2	44.2	40.8	35.7	11.5

TABLE 16 (*Continued*)

County	Population 1956	Percentage rural 1956	Percentage male wage earners earning $1499 or less	Percentage of farms classified as commercial 1956	Percentage males in primary industry	Percentage growth of population by natural increase 1951-56
Newfoundland						
Division No. 1	171,213	40.1	46.5	19.5	22.8	13.6
Division No. 2	23,980	66.5	75.2	40.0	50.3	10.9
Division No. 3	21,675	73.3	68.3	14.4	55.7	13.1
Division No. 4	19,631	72.9	58.2	18.6	42.8	25.5
Division No. 5	35,215	24.2	35.0	69.1	23.1	18.6
Division No. 6	33,738	25.2	32.6	90.0	18.3	14.6
Division No. 7	38,209	82.8	80.1	24.4	51.4	13.2
Division No. 8	40,629	92.0	83.8	38.8	63.4	10.2
Division No. 9	19,970	91.2	81.1	26.7	76.1	20.0
Division No. 10	10,814	100.0	40.9	—	31.9	15.3
Total	415,074	55.4	53.2	21.8	37.3	14.4

Source: Computed from data of *Census of Canada, 1951*, and *1956*.

Brunswick by 13.3 per cent, and Prince Edward Island by 7.3 per cent. Prince Edward Island which is almost entirely dependent on agriculture shows the highest migration rates in the Maritime Region, while Nova Scotia shows the lowest rates.

2. In the case of Nova Scotia, declining employment in the coal mining industry was found to be as important a source of out-migration as the decline of the rural sector. As we shall show below, Nova Scotia would probably have recorded net in-migration in 1951-56, had it not been for the large movements of people out of its coalmining areas.

3. The ability of a province to absorb its displaced rural population in other sectors of its economy is greater the more mature is its economic structure and the higher the degree of its urbanization. Examination of migration patterns within the provinces reveals a movement into the (very few) larger urban centers. Of these, the Halifax-Dartmouth area is the only urban complex in the Maritimes which has been able to attract large numbers of net in-migrants. We found that Nova Scotia was better able to absorb the people leaving its rural areas than was New Brunswick, whiel Prince Edward Island was unable to offer alternate employment.

4. The relationship between high migration rates and low average wages shown on the provincial level in Table 14 was found to hold with surprising consistency on a county basis (see Table 16). Exceptions to the pattern occured where (*a*) an established industrial area with high wages was hit by a decline of basic industry. Such an area will typically show high rates of out-migration despite a high wage level (e.g., Cape Breton County). (*b*) An urban area of in-migration may have a relatively low wage level. In the manufacturing sector the high wage industries are typically the resource-processing industries. These are not usually located in the large cities. Further, the inflow of people to these large cities may depress the wage level.

5. In rural areas heavily dependent on farming there was a strong correlation between high migration rates and low farm income as measured by the percentage of farms classified as commercial.

Prince Edward Island

Prince Edward Island had a virtually stationary population between 1951 and 1956. The rural population declined by 6.7 per cent, while its (much smaller) urban population increased by 23.4 per cent. This increase occurred in small communities. Charlottetown failed to grow by in-migration.

Prince Edward Island is an agricultural province: 45 per cent of all males are employed in agriculture. We made estimates of net migration from the 1951 rural farm population using the survival method. They showed a net out-migration from rural farm population of 6,330 persons in the period 1951-56 equally divided between males and females. Some 70 per cent of these net migrants were under 30 years of age. The farm population aged 15-19 (in 1951) lost 49 per cent of its males and 39 per cent of its females in the five-year period.

Because the survival method understates the out-movement we may (conservatively) adjust our estimate of out-farm movement to 7,000. We know that provincial out-migration from P.E.I. was 8,000 in 1951-56. We further know, that there was also a (much smaller) non-migration from the rural non-farm population. We are safe in concluding that movement out of agriculture accounts for the largest part of migration from P.E.I. Seventy-five per cent of the land of the province is farm land. Between 1951 and 1956 there was a 7.0 per cent decline in the number of farms accompanied by an increase in average farm size. Improved land did not decline. The number of small (10 to 239 acres) holdings declined, the largest drop occurring in the 10- to 69-acres group (16 per cent decline). The number of large holdings increased, the largest increase occurring in the 560- to 759-acre size group (116.7 per cent).

Kings County which is overwhelmingly rural and the poorest in the province produced the highest rate of net out-migration in 1941-51 (21.6 per cent) and the second highest in 1951-56 (8.2 per cent). Prince County is somewhat more prosperous and contains the town of Summerside. It produced the highest rate of net out-migration in 1951-56 (10.7 per cent). This county has the highest rate of natural increase in P.E.I. and two-thirds of the Island's population of French origin are found here. In Queens County we found lower rates of net out-migration in both periods (10.3 per cent in 1941-51 and 6.0 per cent in 1951-56). The reason is the location of Charlottetown in this county. Queens County also showed higher wages (only 52 per cent of male wage earners earning less than $1,500 in 1951) and more prosperous farming (78.5 per cent of farms classified as commercial).

Nova Scotia

Nova Scotia's population increased by 8.1 per cent between 1951 and 1956. Net out-migration was very small: 2,555 males and 8,132 females. Despite continued decline in some sectors of the economy there was an expansion in manufacturing employment of 8.6 per cent.

Although only 13 per cent of Nova Scotia's male population is engaged in agriculture, movement from farm areas was large in the 1951-56 period. The 13.4 per cent decline in the rural farm population of Nova Scotia already referred to was reflected in an estimated net migration of 19,530 persons out of rural farm areas, equally divided between males and females. This estimate is probably high by reason of definitional changes which transferred some 12,000 persons in the Sydney-Glace Bay area from the rural to the urban classification. We thus adjusted the estimate to 18,000. The 1956 census records a 10.4 per cent decline in the number of farms and a 4.8 per cent reduction in acres of improved land, indicating considerable abandonment of farm land. The decline in the number of farms was evenly distributed by size of farm. Nova Scotia thus started to conform to the general Canadian pattern of concentration in farming operations.

We selected those five counties of out-migration which showed the highest percentage of males occupied in farming according to the 1951 census (see Table 17). We found a very strong inverse correlation between the rate of net out-migration from these counties and our index of farm prosperity: thus the counties of out-migration listed below, where livelihood depends primarily on farming, produced 4,584 net out-migrants and an average net migration rate of 5.5 per cent. Six other counties, which together with Halifax, account for the major part of Nova Scotia's fishing industry, produced 7,641 net migrants and recorded an average net migration rate of 6.5 per cent. These counties were Yarmouth 2,073; Digby 1,556; Guysborough 1,476; Shelburne 915; Lunenburg 769; Richmond 742.

TABLE 17

Relation between Net Migration Rates and from Prosperity in Five Leading Agricultural Counties

County	Net migration rate	Percentage of farms classed commercial
Victoria	—8.1	21.5
Inverness	—7.8	36.7
Annapolis	—7.2	43.0
Antigonish	—2.6	48.0
Hants	—2.5	55.1

The principal source of provincial net out-migration from Nova Scotia was not, however the movement out of farm and fishing areas, important as this was, but movement out of counties hearily dependent on coal mining. Fifteen of the eigtheen counties of Nova Scotia showed out-migration. The three exceptions were Kings (+1079), Colchester (+963), and Halifax (+15,060). The remaining 15 counties produced a total net out-migration of 27,289.

We note that over half this number of net out-migrants was generated in the four urban counties of the province.[11] In particular, 14,513 net outmigrants can be attributed to the three counties which contain Nova Scotia's coal mining industry (Cape Breton, 8,963; Cumberland, 2,785; Pictou, 2,765). The depressed state of these three urban counties thus generated a volume of net migrants very considerably in excess of net out-migration from the province. We may conclude that if it were not for the depressed state of the coal mining industry, Nova Scotia

[11] We define a courty as urban where over 50 per cent of the population is classified as urban.

would, in the 1951-56 period have had a net in-movement of people, particularly males. Thus we believe the economy was able to absorb considerable internal population movement.

The picture of population movements in Nova Scotia which emerges is one of historical population decline or very slow growth in eleven of the eighteen counties. These include the fishing ports of the province and some of its farm areas. Cape Breton was the fastest growing county in Nova Scotia until 1941. Out-migration from Cape Breton 1941-51 though large, did not proceed nearly as rapidly as it did in the 1951-56 period. Halifax has shown growth at all times but never at the rate experienced in the early fifties. The picture suggests that rural sources of out-migration were more important in the 1941-51 period than in the 1951-56 period. It presents a process of concentration of population growth around Halifax, reflected in the location of 47 per cent of all males employed in trade, finance, and services in Halifax county in 1951.

In studying the relationship between net migration rates and male wage earners earnings less than $1,500 in 1951 for the eighteen counties of Nova Scotia for the period 1951-56, urban we note that the three counties dependent on coal mining show a *high* rate of net out-migration and a *high* average wage level. If we set these three coal minings counties aside, we may then observe a marked linear correlation between heavy net out-migration and low average wages in the remaining fifteen counties. Our justification for excluding the three coal mining counties from the analysis lies in our hypothesis that the primary source of net out-migration is the relative secular decline of the rural occupations of farming, fishing, and forestry and the consequent decline of areas heavily dependent on primary activity. The correlation coefficient for these fifteen counties was found to be .67.

We note also that *all* counties where 60 per cent of more of male wage earners were earning less than $1,500 in 1951 experienced net out-migration in the period 1951-56 at the rate of 5 per cent or more. These counties were Yarmouth, Victoria, Inverness, Richmond, Shelburne, Digby, and Guysborough.

In the 1941-51 period the relationship between net out-migration and the same index of wage rates was even more marked. Once again, excluding Cape Breton, Pictou, and Cumberland counties, the product-moment correlation coefficient for the remaining fifteen counties of Nova Scotia was .75.

New Brunswick

The population of New Brunswick increased by 7.5 per cent between 1951 and 1956. Net provincial out-migration was 20,700 or 4.0 per cent of 1951 population. The shift by net migration from rural farm population was estimated by us to be 30,000 between 1951 and 1956, resulting in a decline of 13.3 per cent in the rural farm population. We have reason to trust that this estimate is accurate. We note that it represents a considerably larger movement than that observed in Nova Scotia, even allowing for the initially larger rural farm population of New Brunswick. The 1956 census shows a 16.3 per cent decline in the number of farms, a percentage decline greater than any recorded elsewhere in Canada. It represents the disappearance of 4,315 farms in five years. There was a 5.5 per cent decline in improved farm land and a 14.1 per cent decline in total farm land. Concentration of farms by size proceeded in New Brunswick, as in P.E.I.

New Brunswick showed small in-migrations in five of its fifteen counties and substantial out-migration in the remaining ten counties. The small in-migrations are related to the very modest growth of larger Saint John, Moncton, and Fredericton. None of these cities experienced the dynamic growth of Halifax.

With the exception of Westmorland, the ten counties of out-migration are overwhelmingly rural. The percentage of rural population in these nine counties ranges from a low of 54.5 per cent in Restigouche to a high of 90.7 per cent in Queens County. Of these nine counties, seven have more than 40 per cent of males engaged in primary industry, while four have more than 50 per cent of males engaged in primary industry. It is almost impossible to classify the ten counties according to the main economic activity. Charlotte County is heavily dependent on fishing; Carleton, Victoria, and Kent counties are chiefly agricultural; Restigouche and Northumberland are chiefly dependent on the forest although there is some farming. In Gloucester, Madawaska, and Queens counties both farming and forestry are important.

In the counties where farming is important we can again observe in Table 18 the inverse correlation between net out-migration and the prosperity of the farming operation. Victoria County with its extraordinarily high rate of migration and Gloucester County with its extraordinarily low rate of migration form the exception to the pattern.

TABLE 18

Relation Between Net Migration Rates and Farm Property in Six Agricultural Counties

County	Percentage of farms classified as commercial	Net migration rate
Carleton	—6.3	72.3
Victoria	—11.3	69.4
Madawaska	—7.0	52.9
Queens	—9.6	50.6
Kent	—10.0	42.8
Gloucester	—4.4	14.9

Westmorland is the county which contributed the largest out-migration in 1951-56 (—3,618). Although Westmorland County includes Moncton and is thus largely urban, it nevertheless contains a large rural population. Kent and Gloucester Counties both contributed large numbers to the flow of migration, although the migration rate in Gloucester County is lower than the very poor economic circumstances of the area might lead one to expect. So, for that matter, is the rate of out-migration for Northumberland.

Some of the counties producing heavy out-migration since 1941 have showed population decline, or very slow growth, since 1901. They are Kent, Carleton, Queens, and Charlotte. As in Nova Scotia, this is a reflection of secular decline in agriculture and fishing. With the exception of Kent, these counties have low birth rates and are situated in the lower and middle Saint John valley.

The remaining counties of heavy out-migration are quite distinct in their social and economic composition. These counties are Victoria, Madawaska, Restigouche,

Gloucester, and Northumberland. These all experienced rapid population growth since 1901. They all have high birth rates. They include the larger part of New Brunswick's French-Canadian population. Agriculture in the northshore counties is of a subsistence nature. The problem of finding employment for the large surplus of population accumulating within these counties is particular to New Brunswick in the Maritime region, although it is similar to that faced by Newfoundland in many respects. Unlike Newfoundland, however, New Brunswick failed to show any increase in manufacturing employment between 1951 and 1956.

A linear correlation analysis of county net migration rates and our index of county male wage rates for the period 1951-56 for New Brunswick showed a correlation coefficient of .55 for twelve of the fifteen counties of this province. Three counties (Kings, Albert, and Sunbury) were excluded from the analysis because they were heavily affected by suburban growth of the neighbouring urban complexes of Saint John, Moncton, and Oromocto. Again, we found that low wage-level counties where 45 per cent or more of male wage earners were earning less than $1,500 per annum included all the counties of heavy net out-migration. Of these, the following counties experienced net out-migration at the rate of 5 per cent or more over the period 1951-56: Queens, Charlotte, Madawaska, Restigouche, Carleton, Kent, and Victoria. Gloucester and Northumberland counties, although among the poorest counties, showed surprisingly low rates of net out-migration. These low rates of out-movement, despite the poverty of these northshore counties, requires some explanation. One element in such an explanation might be activity in the forest and mining industries in the Bathurst and Newcastle areas. Another element in the low mobility found here is probably to be found in improved social security payments, including seasonal unemployment benefits, which have put a floor under the standard of living.[12] There certainly was a sharp reduction in out-migration rates from both these counties as compared with the earlier 1941-51 period. The net migration rate from Gloucester County 1941-51 was 16.6, whereas for 1951-56 it was only 4.4 per cent. Similar figures for Northumberland county were 11.0 per cent and 3.6 per cent. Restigouche county also showed a dramatic decline in the rate of net out-migration as between the earlier and the later period. The correlation between net migration rates and wage levels in New Brunswick for all fifteen counties in the 1941-51 period (correlation coefficient .66) was higher than in the later period.

Implied in our argument is a strong inverse correlation between dependence of an area on primary industry and a low average wage level. A linear correlation between the percentage of males occupied in primary industry and the percentage of male wage earners earning less than $1,500 in all New Brunswick counties in 1951 is .87. The same relation for all eigtheen counties of Nova Scotia is .57. We once more can note the exceptional case of Cape Breton with the highest wage level of any county in Nova Scotia and a very strong dependence on primary industry. The reason, of course ,is that coal mining differs radically from rural primary activity and is associated with urbanization and a high wage level. If we

---

[12] In so far as increased social payments have reduced geographic mobility, this marks one of the achievements of the welfare state. As such it is both desirable and essentially irreversible. The days when people literally could be starved out in the interests of labour mobility are gone. Thus the need arises to create the conditions under which the differential between subsistence and earned income is sufficiently large to draw people into active employment.

exclude the three urban counties of Cape Breton, Pictou, and Cumberland, where there is heavy dependence on coal, we obtain a correlation coefficient of .76 for the remaining fifteen counties of Nova Scotia.

These relationships lend substance to the popular belief that an area will be richer the less is its dependence on primary industry and the larger is its manufacturing sector and the greater the extent of urbanization. Thus regional differences between family incomes of metropolitan families in Canada are far smaller than regional differences among (non-farm) family incomes of non-metropolitan families. It is also a fact that regional differences in income from full-scale farms are much less than differences indicated by regional average farm income figures. Both sets of facts point towards the underdeveloped structure of the economy of the Atlantic provinces, with its large subsistence sector of low-income agriculture, combined with part-time fishing and part-time logging as the cause of differentially lower incomes. In so far as even the large movements of population described

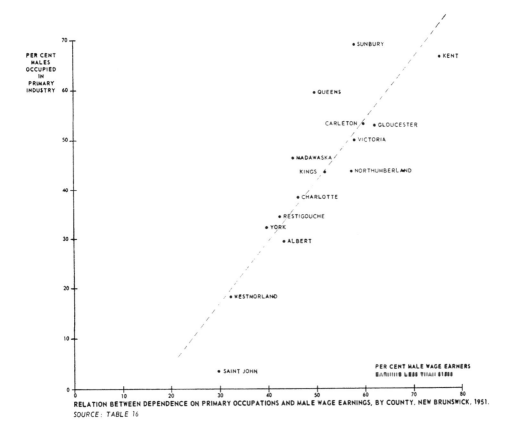

RELATION BETWEEN DEPENDENCE ON PRIMARY OCCUPATIONS AND MALE WAGE EARNINGS, BY COUNTY, NEW BRUNSWICK, 1951.
SOURCE: TABLE 16

in this paper have failed to dissolve the rural surplus, especially in areas with high birth rates, such as the northern counties of New Brunswick, it is natural that many people in the area look to accelerated industrial development as the solution to the economic problem of the area. Demographic trends towards a higher rate of entry into the labour force lend urgency to this belief.

MIGRATION, EMPLOYMENT, AND WAGES

We have already referred to the relation between county net migration rates and the percentage of persons earning less than $1,500 per year in 1951. Further relations between migration and wage levels are presented in Table 19, where we have lined up net migration rates with the average county wage and salary earned in *manufacturing* industries averaged over the years 1951-56. We were also curious to see whether there exists any firm relation between migration rates, the *growth* of employment, and *changes* in average wages. Data on the growth of total employment would have been more pertinent. Such data were not, however, available on a county basis (or indeed on a provincial basis). We thus used data on employment and wages and salaries in the manufacturing industries.

Nova Scotia

Manufacturing accounts for 18 per cent of provincial employment in Nova Scotia, 15 per cent in New Brunswick, and 10 per cent in Newfoundland.

In Nova Scotia, manufacturing employment expanded by 9 per cent (1951-56), while the average wage-salary increased by 40 per cent over the same period to a level of $2,400 in 1956. We note that the two counties with average wage-salary above provincial average level both showed a high rate of out-migration. This is contrary to the general correlation between high migration rates with low wages. The reason is obvious: Cape Breton ($3,022) and Pictou ($2,517) are both urban-industrial areas with a general wage level far above that prevailing in rural areas. The heavy out-migration from these counties is due to the decline in the coal mining and primary iron and steel industries. The same is true of Cumberland County which also has a relatively high manufacturing wage rate ($2,091). This again emphasizes the important distinction which must be made between migration from rural-subsistence-activity origin and that of urban-industrial origin.

If we remove these three urban countes, every other county of out-migration in Nova Scotia has average wage-salary below the provincial level, while Halifax, the sole county of heavy in-migration, shows an average wage-salary level in excess of the provincial level ($2,619). The average wage-salary in manufacturing in Halifax rose by 44 per cent (1950-56) and the volume of manufacturing employment also rose faster than the provincial average (21 per cent).

In ten of the fifteen counties of Nova Scotia for which we have data above-average growth in employment was associated with above-average *increase* in wage-salary level, and vice versa. There thus appears to be a strong relation between growth in manufacturing employment and rise in manufacturing wages.

New Brunswick

New Brunswick presents a different picture. Manufacturing employment contracted by 5.5 per cent (1950-56). The average wage-salary level, however, rose by almost as much as that of Nova Scotia (38 per cent) and stood fractionally higher in 1956 at $2,450. The reason lies in the different structure of the manufacturing industry of New Brunswick and, in particular, in the large high-wage pulp and paper industry. Thus five of the eight rural counties of out-migration in New Brunswick, for which we have data, experienced an *increase* in manufac-

TABLE 19

Net Migration Rates, Growth of Manufacturing Employment and Average Wage and Salary
in Manufacturing Industry; Counties of the Atlantic Provinces,
1951-1956

	Net migra-tion rate 1951-56	Manufacturing employment, 1956 (1950=100)	Average manufacturing wage and salary, 1956 (1950=100)	Average manufacturing wage and salary, 1951-56 $	Percentage of all wage earners earning less than $1,500
		Nova Scotia			
Heavy out-migration					
Guysborough	—10.4	122*	144*	1,400	70.6
Yarmouth	— 9.1	77	127	1,783	60.6
Victoria	— 8.1	125*	136	1,106	61.4
Inverness	— 7.8	105	116	910	64.2
Digby	— 7.8	78	116	1,327	72.9
Cape Breton	— 7.4	98	138	3,022*	18.8*
Cumberland	— 7.0	111*	149*	2,091	46.0
Richmond	— 6.9	664*	106	1,456	63.3
Pictou	— 6.3	133*	139	2,517*	37.7
Shelburne	— 6.3	95	128	1,517	70.3
Queens	— 5.3	—	—	—	—
Antigonish	— 2.6	—	—	—	—
Hants	— 2,5	108	142*	1,918	56.9
Lunenburg	— 2.3	—	—	—	—
Mixed migration					
Annapolis	— 7.2	83	138	1,670	37.3*
Colchester	+ 1.4	106	132	1,778	47.3
Kings	+ 3.3	96	147*	1,706	57.1*
In-migration					
Halifax	+ 9.3	121*	144*	2,619*	27.0*
Nova Scotia	— 1.7	109	140	2,400	38.0
		New Brunswick			
Heavy out-migration					
Victoria	—11.6	106.2*	142.0*	2,061	59.8
Kent	—10.0	100.0*	153.0*	1,118	75.4
Queens	— 9.6	—	—	—	—
Charlotte	— 9.2	73.0	127.0	1,711	46.1
Madawaska	— 7,0	118,6*	131,0	3,319*	45,1
Carleton	— 6.3	96.0	137.0	1,615	59.6
Restigouche	— 5.7	113.6*	140.0*	3,529*	42.4*
Gloucester	— 4.4	113.0*	154.0*	2,520*	62.0
Northumberland	— 3.6	79.9	153.0*	2,254	57.3
Mixed migration					
Westmorland	— 4.5	86.4	142.0*	2,549*	32.1*
Saint John	+ 0.3	104.8*	130.0	2,496*	29.7*
York	+ 1.0	7.4	123.0	1,952	39.8
New Brunswick	— 4.0	94.5	138.0	$2,450	44.2

TABLE 19 (*continued*)

	Net migration rate 1951-56	Manufacturing employment, 1956 (1950=100)	Average manufacturing wage and salary, 1956 (1950=100)	Average manufacturing wage and salary, 1951-56 $	Percentage of all wage earners earning less than $1,500
			Newfoundland		
Census Division No. 3	— 7.0	115*	127*	1,356	68.2
,, 7	— 4.9	98	77	916	81.1
,, 2	— 3.7	109	130	1,534	75.2
,, 9	— 2.8	91	91	954	81.1
,, 4	— 2.7	127*	119	832	58.2
,, 8	+ 0.2	98	140*	714	83.8
,, 1	+ 0.9	120*	131*	2,079	46.5*
,, 6	+ 6.0	98	142*	4,145*	32.6*
,, 5	+ 6.8	108	116	4,033*	35.0*
,, 10	+21.8	—			
Newfoundland	+ 0.5	109	123	2,600	53.2

Source: *The Manufacturing Industries of Canada: Regional Distribution* (D.B.S.), various issues.

* Denotes above provincial average.

turing employment in 1950-56 while, as we mentioned already, provincial manufacturing employment declined. In four of these counties (Victoria, Restigouche, Kent, and Gloucester) manufacturing wage-salary rose by more than the provincial average, whereas in the fifth case (Northumberland) average wages also rose by more than the provincial percentage, despite a decline in manufacturing employment. Thus in Restigouche ($3,529), Madawaska ($3,319) and Gloucester counties ($2,520) we found an average manufacturing wage-salary level higher than the provincial level and higher than average wages in Saint John. Yet these three counties all produced out-migrations at above-average rates. Gloucester county is one of the poorest areas in the province. We thus note some manifestation of a dual economy here where a high-wage manufacturing sector based on natural resources coexists with severe poverty and is unable to generate sufficient secondary employment in the area to absorb the rural labour surplus.

Westmorland County which produced a high rate of out-migration is also a high-wage urban area ($2,549). In 1941-51 this county had considerably in-migration. Out-migration from Westmorland in the 1951-56 period has thus the same character as migration from the urban areas of Nova Scotia already discussed.

While manufacturing employment in Saint John did increase somewhat, the wage level failed to rise as fast as the provincial average. Average wages in Saint John ($2,496) in 1956 were lower than wages either in Westmorland or in the rural counties where the paper industry is sited. Saint John thus presents a depressing contrast to the dynamic growth of manufacturing employment and rise in wage level found in Halifax. York County, which contains Fredericton, was also an area of in-migration. It showed a decline in manufacturing employment equalled only in Charlotte County, a very modest rise in wages, and a very low level of average wage-salary in 1956 ($1,952). This confirms our previous impression of structural underdevelopment in New Brunswick where the growth of the pulp and paper industry has failed to generate secondary manufacturing either in the cities or in the countryside on a scale sufficient to employ the rural surplus. In general, the wage gap seems to widen, with the largest increases occurring in these areas where the high-wage activities are located *despite* large-scale under-employment in these areas.

Newfoundland

In Newfoundland manufacturing employment expanded by 23 per cent. Although Newfoundland has much lower per capita incomes than either New Brunswick or Nova Scotia, the provincial average wage-salary in 1956 was $2,600, higher than either of these Maritime provinces. The reason is again the pulp and paper industry which accounts for one third of manufacturing employment in Newfoundland. The geographical areas which contain Corner Brook and Grand Falls showed average annual wage-salary levels of $4,033 and $4,145, respectively, the highest manufacturing wage-salary on a county basis in the entire Atlantic region and twice as high as the average manufacturing wage in St. John's ($2,079). Unlike the forest areas of New Brunswick, however, Corner Brook and Grand Falls attracted large in-migrations (6.8 per cent and 6.0 per cent).

Out-migration in Newfoundland thus flowed from the areas with the lowest average wages to those with high average wages, although, as we already mentioned, St. John's which received modest in-migration, did not reach the provincial wage level in 1956. The situation differs from that of New Brunswick in that St. John's experienced a 20 per cent increase in manufacturing employment (1951-56) and an increase in average manufacturing wage-salary of 31 per cent, as compared with a provincial increase of only 23 per cent. Such growth in manufacturing employment for the local market which took place on Newfoundland is heavily concentrated in and around St. John's. Thus 82 per cent of male employment in manufacturing other than pulp and paper, fish-processing and sawmills is found in the Avalon Peninsula. As in the case of Halifax, trade, finance and service employment is also heavily concentrated in the capital (60 per cent of male employment).

The area of heaviest out-migration is the south shore (Census Division # 3) where assisted relocation has encouraged migration from isolated outports. Manufacturing employment (fish-processing) in this area has increased, the average wage-salary (27 per cent) has also increased and was $1,356 in 1956. A similar development is found in Census Division # 2, where average wage was $1,534. Other areas of out-migration such as Census Division $ 7 (Bonavista Bay and the north shore of Trinity Bay); Census Division # 9 (northern peninsula); and Census Division # 4 (Stephenville and St. George Bay) show the lowest average manufacturing wages found anywhere in the Atlantic provinces ($916, $954 and $832, respectively). Furthermore, average wages here have actually *declined* since 1951. By contrast, the average wages in Grand Falls, for instance, which in 1956 was $4,145 and the highest in the Atlantic region *rose* by 42 per cent in the period 1951-56. The process of economic development in Newfoundland has thus acted to widen the gap between rich and poor areas and to hasten concentration of secondary activities in the single important urban agglomeration of St. John's. On the South Shore, out-migration has been accompanied by a healthy rise in employment and in the wage level. This however is the area where the government has intervened to assist in rationalization, relocation, and employment development. In other areas dependent on fishing, out-migration has proceeded very slowly and has not prevented the deterioration of the volume of employment and earnings. The contrast of the communities with those of the two booming pulp and paper developments is depressing. It thus appears that the market growth forces associated with the paper industry are insufficient, by themselves, to bring about improvement in the old established sectors of the economy on which the great majority of the population depend for their livelihood.

APPENDIX ON STATISTICAL METHOD

The Natural Increase Method

For the best estimate, natural increase should be calculated by census year (June 1 to May 31) rather than by calendar year. For this reason the estimates presented in Table A1 differ slightly from those shown in Bulletin 3-1 of the *Census of Canada, 1956*. The difference is very small.

No method of estimating is infallible. The accuracy of this method depends on accurate vital statistics. Where a considerable number of babies are not born in hospitals, as is still the case in Newfoundland, it is probable that births are understated more than deaths. In this case net in-migration would be *over*stated by the natural increase method and net out-migration *under*stated. The method also depends on accurate census data. It is probable that the census under-enumerates the population, especially small children. If a population is growing very fast and thus has a high proportion of young children, such errors, even though common to the Census at both dates, would work in the *opposite* direction: net in-migration would be *under*stated and net out-migration would be *over*stated. Where either of these sources of error may be present as is the case for instance, in Newfoundland and, where, moreover, net migration is very small as it was here in 1951-56, we cannot be really sure whether there was, in fact, in-migration to Newfoundland.

The natural increase method is not confined to measuring interprovincial net migration. D.B.S. has published net in- and out-migration for counties and census subdivisions and for metropolitan areas. Such estimates can be made for any city, town, village or township, provided boundaries do not change between census dates.

Survival Method

The natural increase method can be used only where vital data are available. Thus, it cannot be used before 1921; nor can it be used to obtain estimates of net migration for categories of the population other than geographical ones. It *can* be used to estimate the age-distribution of net migrants, but the process is tedious and time-consuming. For these reasons we have made extensive use of the survival method. This method was developed by Dr. Nathan Keyfitz when he was with the Dominion Bureau of Statistics and used by him to make historical estimates

TABLE A1

Estimated Net Migration by Natural Increase Method for Census Years 1951-1956, by Sex

	Newfound-land	Prince Edward Island	Nova Scotia	New Brunswick
Males				
Population 1951	185,143	50,218	324,955	259,211
Natural increase 1951-56	25,835	4,389	31,111	29,445
Expected population 1956	210,978	54,607	356,066	288,656
Census population 1956	213,905	50,510	353,182	279,590
Net migration	2,927	—4,097	—2,884	—9,066
Females				
Population 1951	176,273	48,211	317,629	256,486
Natural increase 1951-56	26,016	4,570	32,022	30,329
Expected population 1956	202,289	52,781	349,651	286,815
Census population 1956	2101,169	48,775	341,535	275,026
Net migration	—1,120	—4,006	—8,116	—11,789
Total net migration	1,807	—8,103	—11,000	—20,855

of net migration into and out of Canada, as well as interprovincial migration.[13] We calculate the prohability of surviving five years, let us say, for each age group. More exactly, for the age group 5-9, we calculate the probability that a male aged seven will survive to age twelve. This probability is the product (p_{-8} x p_{-9} x p_{-10} x p_{-11} x p_{-12}) where p_{-8} represents the probability that a male aged seven will survive for one year, i.e., until he is eight. These probabilities of surviving one year are presented in life tables and are based on age specific death rates. The survival rates which were used in this study are reproduced in Table A2, and Table A3 uses these to estimate net migration.

Estimate of Migrants Aged 0-4

Such an estimate must be added to the estimate of migrants 5 + obtained by the survival method in order to permit comparison with estimates derived by natural increase method.

(a) *Births* between mid-1951 and mid-1956 are summed from annual vital statistics, for each province. (b) *Deaths among children* who would be agel 0-4 in June 1956 (0-4) are summed as in Table A4. (c) *Natural Increase*, (0-4) is obtained by subtracting deaths (0-4) from births (see Table A5). (d) *Net Migration* (0-4) is obtained by subtraiting census population. The estimates made by the two methods are compared in Table A7.

The survival method consistently underestimates out-migration and over-estimates in-migration. Two reasons are offered to account for this:

1. *Mortality.* We used the Canadian Life Table, 1951, which was based on mortality as found in 1950-52. In fact, age-specific death rates declined during the intercensal period 1951-1956. Furthermore, Canadian mortality rates were

[13] See the "Growth of Canadian Population," *Population Studies,* vol. IX, no. 1, June, 1950.

TABLE A2

Survival Rates Used to Calculate Expected Populations

	1941 survival rates probability of living ten years		1951 survival rates probability of living five years	
	Maritime males	Maritime females	Canada males	Canada females
5-Year age groups				
0-4	.978309	.980172	.993476	.994900
5-9	.984755	.987343	.995840	.997150
10-14	.978775	.982134	.994964	.996901
15-19	.971216	.974057	.991998	.995652
20-24	.968075	.969741	.990817	.994821
25-29	.964237	.966709	.990753	.993850
30-34	.958396	.958639	.989080	.991732
35-39	.949965	.950594	.984878	.988082
40-44	.933107	.940653	.976155	.982212
45-49	.903328	.918970	.961523	.973998
50-54	.858674	.887125	.939377	.961903
55-59	.794557	.839570	.907810	.941194
60-64	.702295	.759573	.866875	.909127
65-69	.572679	.639491	.810058	.857393
70-74	.411751	.476039	.718861	.770804
75-79	.246417	.302844	.589953	.643474
80-84	.116236	.163944	.438439	.484330
85-89	.044731	.041189	.285760	.315142
90-94	.012540	.016630	.155321	.165921
95	.002064	.000500	.062953	.062901
10-Year age groups				
25-34	.961285	.963109	—	—
35-44	.942831	.946791	—	—
45-54	.882715	.903845	—	—
55-64	.752104	.804067	—	—
70	.311343	.360000	—	—

Source: *Life Tables for Canada & Regions 1941*, Vital Statistics Analytical Report #4. *Canadian Life Tables 1951*, D.B.S. Reference Paper #50.

generally higher than those prevailing in the Atlantic provinces. Table A 8 illustrates both the fall in mortality rates over the period and the fact that standardized death rates in Nova Scotia and Prince Edward Island were considerably lower than the Canadian average. Mortality in Newfoundland and New Brunswick was also somewhat lower than Canadian mortality. Thus the number of persons who would actually have died in each of the Maritime provinces was smaller than our estimates show. Our expected population is therefore too small, and our estimates of net out-migration too low, while those of net in-migration are too high.

2. *Under-Enumeration of Small Children.* Another source of error in the survival method is the evident under-enumeration of small children in the census. The result of under-enumerating children 0-4 in 1951 is to *underestimate* out-mi-

TABLE A3

Estimated Net Migration for Atlantic Provinces, 1951-1956 by Survival Method, by 5-Year Age Groups for Males and Females

The Example of Prince Edward Island

Age group	Population 1951 census		Population expected 1956		Population 1956 census		Net Migration 1951-1956		Net migration rate	
	M	F	M	F	M	F	M	F	M	F
0-4	6,705	6,508	6,681§	6,394§	6,322	5,963	—359§	—431§	—	—
5-9	5,302	5,056	6,661	6,475	6,393	6,128	—268	—347	—4.00%	—5.34%
10-14	4,778	4,516	5,280	5,042	5,031	4,797	—249	—245	—4.70%	—4.85%
15-19	4,176	4,120	4,754	4,502	4,115	4,075	—639	—427	—13.38%	—9.46%
20-24	3,345	3,212	4,143	4,102	3,115	3,032	—1028	—1070	—24.62%	—25.98%
25-29	3,214	3,241	3,314	3,195	2,721	2,648	—593	—547	—17.73%	—17.04%
30-34	3,176	3,108	3,184	3,221	2,785	2,848	—393	—373	—12.42%	—11.51%
35-39	3,223	3,067	3,141	3,082	2,963	2,932	—178	—150	—5.61%	—4.83%
40-44	2,862	2,489	3,174	3,030	2,903	2,890	—271	—140	—8.41%	—4.57%
45-49	2,366	2,120	2,794	2,445	2,765	2,318	—29	—127	—1.02%	—5.11%
50-54	2,368	2,131	2,275	2,065	2,230	2,022	—45	—43	—1.91%	—2.03%
55-59	2,097	2,033	2,224	2,050	2,140	1,945	—84	—105	—3.55%	—4.93%
60-64	1,793	1,716	1,904	1,913	1,963	1,891	59	—22	2.81%	—1.09%
65-69	1,666	1,602	1,554	1,560	1,703	1,622	149	62	8.30%	3.61%
70	3,147	3,292	3,192	3,416	3,361	3,664	169	248	3.51%	5.06%
Total	50,218	48,211	47,594*	46,098*	44,188†	42,812†	—3,406‡	—3,286‡	—6.79%	—6.82%

* Expected 1956 population aged 5 and over.

† 1956 census population aged 5 and over.

‡ Net migration 1951-56 from the 1951 census population (sometimes referred to as migrants 5+).

§ Estimate of migrants 0-4 arrived at by the calculation set out below.

KARI LEVITT

TABLE A4

0-4 Deaths during the Period 1951-1956

Deaths 0-4 for Census Years for 1951-1956 among population who would be age 0-4 on June 1, 1956

	Age	New-found-land	Prince Edward Island	Nova Scotia	New Bruns-wick
Males					
1951-1952	0	288	41	301	376
1951-1952	0	293	48	313	382
	1	63	11	56	67
1952-1953	0	289	33	293	333
	1	64	6	58	59
	2	15	1	16	16
1953-1954	0	306	50	314	337
	1	59	8	70	60
	2	17	2	18	18
	3	10	2	11	14
1954-1955	0	294	50	284	311
	1	57	6	49	57
	2	17	4	16	20
	3	12	3	10	19
	4	9	1	6	14
Totals		1,793	266	1,815	2,083
Females					
1951-1952	0	225	20	223	281
1952-1953	0	219	34	215	273
	1	50	12	36	57
1953-1954	0	201	28	184	248
	1	44	6	40	58
	2	9	—	11	14
1954-1955	0	214	43	198	253
	1	44	7	55	38
	2	12	2	12	15
	3	6	—	7	12
1955-1956	0	230	37	200	212
	1	52	4	32	57
	2	14	1	12	14
	3	10	1	7	12
	4	7	—	5	9
Totals		1,338	195	1,237	1,553

* Age 0. Taken from month of death by month of birth Tables, by provinces and sex.

TABLE A5

Natural Increase, 0-4 Years, Atlantic Provinces, 1951-1956

	New-foundland	Prince Edward Island	Nova Scotia	New Brunswick
Males				
Births	33,937	6,947	47,309	42,197
Deaths	—1,793	—266	—1,815	—2,083
Natural Increase	32,144	6,681	45,494	40,114
Females				
Births	32,453	6,589	44,559	40,205
Deaths	—1,338	—195	—1,237	1,553
Natural Increase	31,115	6,394	43,322	38,652

TABLE A6

Net Migration, 0-4 Years, Atlantic Provinces, 1951-1956

	Expected Population 0-4 in 1956		Census Population 0-4 in 1956		Net Migration (0-4) 1951-1956	
	Males	Females	Males	Females	Males	Females
Newfoundland	32,144	31,115	31,933	31,441	—211	326
Prince Edward Island	6,681	6,394	6,322	5,963	—359	—431
Nova Scotia	45,494	43,322	44,027	41,945	—1,467	—1,377
New Brunswick	40,114	38,652	37,713	36,586	—2,401	—2,066

TABLE A7

Comparison of Estimates of Net Migration Made by Survival Method and by Natural Increase Method for June 1951-June 1956

	Male	Female	Total
Newfoundland			
By survival			
Migrants 5+	3,852	—1,325	
Migrants 0-4	—211	326	
Total	3,614	—999	2,615
By natural increase	2,927	—1,120	1,807
Prince Edward Island			
By survival			
Migrants 5+	—3,406	—3,286	
Migrants 0-4	—359	—431	
Total	—3,765	—3,717	—7,482
By natural increase	—4,097	—4,006	—8,103

TABLE A7 (*continued*)

	Male	Female	Total
Nova Scotia			
By survival			
Migrants 5+	—536	—5,761	
Migrants 0-4	—1,467	—1,377	
Total	—2,003	—7,138	—9,141
By natural increase	—2,884	—8,116	—11,000
New Brunswick			
By survival			
Migrants 5+	—6,104	—9,275	
Migrants 0-4	—2,401	—2,066	
Total	—8,505	—11,341	—19,846
By natural increase	—9,066	—12,789	—20,855

gration among children 5-9 between 1951-1956.[14] If we compare net migration among children aged 0-4, as obtained from vital statistics data with estimated net out-migration of children aged 5-9 as shown in Table 5, we find a large discrepancy in Nova Scotia, New Brunswick, and Newfoundland.

TABLE A8

Standardized (Age-adjusted) Death Rates, 1951-1956

	Canada	New-foundland	Prince Island Edward	Nova Scotia	New Brunswick
1951	9.0	8.9	7.4	8.3	9.3
1952	8.8	8.0	7.4	8.1	8.7
1953	8.6	7.6	7.3	8.0	8.6
1954	8.2	8.0	7.5	7.7	7.8
1955	8.2	8.8	6.8	7.9	7.9
1956	8.2	8.3	7.3	7.5	8.1

Source: *Vital Statistics, 1958,* Table 26.

[14] The *survival* method consistently *underestimates* out-migration among the youngest group of net migrants becaus in this group we are comparing a mortality-adjusted enumeration of the group 0-4 with a subsequent census enumeration of the group 5-9. As a result the summed estimate of net migration over all age-groups obtained by the survival method is consistently low in estimating out-migration. This observation does *not* contradict the easter argument in this Appendix which states that the *natural increase* method, may, in a rapidly-growing population result in *over-estimating* net out-migration. The reasoning in this latter case is as follows: the difference between the first enumeration of the whole population plus births less deaths and the subsequent enumeration of the whole population will *not* be entirely due to out migration if the absolute number of children under five has increased between the two censuses and thus the absolute amount of under-enumeration has also increased. In so far as the quality of the census improves (i.e., there is relatively less under-enumeration) the suggested effect will be cancelled out.

A Note on Estimates of Age-Specific Migration Rate

The age distribution of migrants as derived in this study is, we feel, one of the more useful results we obtained. Yet it must be treated with some caution, particularly in the high age groups, where the apparent in-migration of people most likely results from our use of mortality rates which are higher than actual mortality for these people at this time. In other words, our "expected" populations at high age groups too low. What appears as in-migration at these ages is actually a differential in mortality as between all-Canada in 1951 and the Atlantic provinces over the period 1951 to 1956.

This interpretation is substantiated by a check made on the female population in Nova Scotia aged sixty-five in 1951, numbering 27,703. We subtracted the number of deaths by single years of age between June 1, 1951 and June 1, 1956 among these people (using vital data) land obtained a figure of 20,227, which is 21 greater than the female population aged seventy in 1956. The figure of net in-migration we obtained by the survival method for females in Nova Scotia between 1951 and 1956 was 665.

As an independent check on migration of older persons into the Atlantic region, change of address data of persons in receipt of old age pensions were examined. These records were not available for the whole period 1951-56, but only for 1953-55. There was a slight net out-migration of these old age pension recipients for Nova Scotia, New Brunswick, and Prince Edward Island, with Newfoundland showing a slight net in-migration.

Use of Survival Method to Estimate Net Migration from Cultural Groupings: An Example

The method used to estimate net migration 1941-51 by origin, language and religious characteristics is illustrated in Table A 9, by presenting the calculation for British male migrants from Prince Edward Island underlying Table 8. A similar calculation was made for French males and for all remaining (other) males. The results are also presented in Table 8.

Census Definitions, 1951

Origin. "For census purposes a person's origin or cultural group was traced through the father. Whenever applicable the origin of a person was established by asking the language spoken by the person, or the paternal ancestor on first arrival to the continent".

Mother Tongue. "By mother tongue is meant the language the person first spoke in childhood and still understands. In the case of migrants this refers to the language commonly spoken in the home."

Religious Denomination. "The specific religious denomination of which the person was either a member or to which he adhered or favoured was enumerated."

KARI LEVITT

TABLE A9

Net Migration, 1941-1951, British Male Migrants, Prince Edward Island

Age group	Population of British origin 1941 census	Probability of surviving 10 years	Population of British origin expected in 1951	Population of British origin 1951 census	Estimated net migration (1941-1951)
0-4	3819	.978309	—	—	—
5-9	3907	,984755	—	—	—
10-14	3869	.978775	3736	3794	58
15-19	3841	.971216	3847	3374	—473
20-24	3722	.968075	3787	2670	—1117
25-34	6063	.961285	7333	5303	—2030
35-44	4382	.942831	5828	5047	—781
45-54	4031	.882715	4132	3987	—144
55-64	3214	.752104	3558	3331	—227
65-69	1308	.572679	3967*	4098*	131*
70	2574	.311343			
Total	—	—	36187†	31604†	—4583‡

* Expected and actual population in 1951 aged 65 and net migrants 65.

† Expected and census population 10+, 1951.

‡ Net migration 10+.

TABLE A10

Method of Calculating Expected Population, 1951

Age group	1941 census population	Probability of survival	Expected population 1951
0-4	P_1	e_1	—
5-9	P_2	e_2	—
10-14	P_3	e_3	P_1e_1
15-19	P_4	e_4	P_2e_2
20-24	P_5	e_5	P_3e_3
25-34	P_6	e_6	$P_4e_4 + P_5e_5$
35-44	P_7	e_7	P_6e_6
45-54	P_8	e_8	P_7e_7
55-64	P_9	e_9	P_8e_8
65-69	P_{10}	e_{10} $\left.\vphantom{\begin{matrix}a\\b\end{matrix}}\right\}$	$P_9e_9 + P_{10}e_{10} + P_{11}e_{11}$
70+	P_{11}	e_{11}	

Discussion

Z. W. SAMETZ

1. MRS. LEVITT'S MODEL concentrates on the net intercensal migration. This is a useful operation for indicating areas of net loss or gain. As she indicates, however, gross migration figures would have been more valuable for most of the characteristics studied. I recognize that a full measure of in-migration and out-migration is not available, but I wish to propose that some further measurements should be made in this direction from (*a*) more intensive census tabulations by area of origin, and (*b*) from annual series which can be developed.

(*a*) The larger policy issues require knowledge of the characteristics of the changed labour supply and its physical mobility pattern. This knowledge must be obtained from the census characteristics of the known new residents and of the known former residents.

For example, Nova Scotia lost a net population of 11,000 in 1951-56. But, by educational criteria, it might actually have gained *qualitatively*. Let us assume this could be a result of gaining 20,000 in-migrants with high-school levels of training (say a mean level of 12 years) and losing 31,000 out-migrants with a mean level of 6 years, so that the educational migration formula $Me = (Ie - Oe)$ would yield $(20,000 \times 12) - (31,000 \times 6)$ or a net gain of 54,000 school years. There are many underdeveloped regions which would not complain of such an exchange. The results of such characteristics are far different if the educational levels were equal, or if the out-migrants had higher skill levels than the in-migrants.

Thus, we might not be able to get the full measure of gross in- (I) and out- (O) migration due to short-term movements within the intercensal period terminating in return or death, but we should have a far better picture of mobility and the characteristics of the migrants. It does make a difference for long-term analysis whether a net loss of 11,000 is the result of (O) $= 11,000$ and (I) $= 0$, or of (O) $= 61,000$ and (I) $= 50,000$ because the two cases have different implications as to basic causes, stability, turnover rate, velocity of circulation, and return potential. It also makes a difference to the analysis if we further know the basic characteristics of the migrants by sex, age, education, birth place, and origin. The language and religious characteristics are less reliable because there are loss or conversion possibilities.

(*b*) On the annual basis, we could try to correlate three partial series on movements, the way Frank Denton did in his paper with four series on employment. The three partial series I suggest can be used are:

C.P.S.A. Conference on Statistics, 1961, *Papers*. Printed in the Netherlands.

(i) the Family Allowance data on movement of families and children (based on the Kasahara paper in this volume),

(ii) the U.I.C. data on the movement of insured workers (based on the Greenway and Wheatby paper in this volume), especially if the single workers can be extracted;

(iii) the movements of tax payers between Taxation Districts, especially on the uninsured single workers.

With these series, we would not only have a closer approximation to the gross movements between census dates, but also an indication of intervening annual movements. This inflow-outflow analysis should give a truer picture of migration looked at from the dynamic aspect, than does the net migration data after cancellation of effects between census dates.

2. The next major comment I would make is that it is unfortunate that the net migration data are not put in the perspective of regional population pyramids to help to explain changes in rates of movement through time. A very high rate of outflow in one decade may reflect itself in a low birth volume in the next decade and a drying-up of out-migration in the succeeding decades. For example, in Table 8, there was a decline in the three age groups within the 15-29 brackets in 1956 compared to 1941. This in itself was a product of the low birth rate of the thirties, which gap in the labour force was partially filled during the 1950's by foreign immigration into other parts of Canada, but much less so in the Maritimes. The lack of gap filler is going to have quite an effect on migration patterns from the Maritimes. It may explain the relatively low levels experienced in the latter fifties, and perhaps if uncorrected, low levels again twenty years from now.

This may be a partial explanation of some abandonment of farms in the Maritimes, and of the continued immigration flow (even if at low absolute levels) into the Maritimes when the immigration rate into the rest of Canada slowed down considerably. There may be an outflow of miners but an inflow of farmers.

3. It is unfortunate that Mrs. Levitt did not feel in a position to offer any explanation for the cultural differences in the mobility rates of the young migrant. There is parallelism in three series which should give sufficient clues to look for an explanation in an intermediate variable such as the high average family size of various groups, or even better the number of children born per family. The migration in particular counties stems not singly from the origin, or the language spoken, or the religion, or because they live on submarginal land, but through their joint product as represented by population pressure within the family, high enough to create both a higher regional population and a heavy outflow of migrants.

4. The most fascinating part of the study to me was the educational attainment section, in view of the current emphasis on training in skills for the emerging economy.

The hypothesis of a greater propensity on the part of the better educated Maritimers to move (all other factors being equal) could not be tested—this is probably why it can continue to remain a popular preconception, a rationalization to explain away the problem. It would however be interesting to compare the educational level of the expatriate Maritimers with that of those who stay at home. I hope that with the new computer facilities, the 1961 census will give us the cross tabulations by province of residence, by province of birth, by the improved educational scaling in the new census.

The evidence in the paper appears overwhelming that most of the emigration is from rural areas (providing seven-eights of the migrants in some provinces) with low educational attainment. It can thus be estimated that a sizable proportion of the out-migrants are not highly educated, probably with lower levels than the average of the areas to which they go.

Further exploration of this field through intensive census analysis may suggest quite another policy recommendation than either the subsidized export thesis of the Gordon Commission or the subsidized industrialization thesis of the A.P.E.C., namely, higher educational training, especially in technical skills.

5. The notes on abandonment of farms do not support the displaced rural population thesis. In Nova Scotia, 4.8 per cent of the improved land was abandoned between 1951 and 1956. In New Brunswick, 5.5 per cent of the improved land (14.1 per cent of the total farm land) was abandoned in those five years. This rural population is not displaced, it is actually creating vacuums. This is not only a movement out of agricultural families—it is a movement out of agriculture.

6. I would even go further, on the basis of this and other evidence, to challenge, for polemic purposes, the basic conclusion that on the whole the pattern of migration conforms to economic incentives. This is not to say that I am setting aside the significant correlation coefficients and regresson coefficients between net out-migration and income that have been worked out for the counties of New Brunswick and Nova Scotia for the two periods 1941-51 and 1951-56. I am however mindful of the wartime attempts to increase coal production by raising wages, which had however the contrary effect of decreasing output. At the higher wage rates, the miners could earn the same income with fewer days per week, and they chose to exercise the option in terms of constant income and fewer hours of work rather than the same or more hours for a higher income.

Thus, I would turn the thesis on its head and say that economic incentives undoubtedly play some role, but that on the whole, "home non economicus" is all too evident. Some of the slowest-growing areas, in fact "closed out areas," show the greatest resistance to movement, as Gerald Fortin's paper suggested with respect to submarginal agricultural zones. Other fast-growing or richly productive areas show excessive movement. The anomalous vacuums are now being filled by immigrant farmers who are establishing very profitable farms, with average net incomes of over $7,000 after initial establishment, supplying milk and meat to the growing Halifax market, to fill an essential local need that would otherwise have to be filled at high prices from long distances. They have the "know-how." Similarly, new agriculture is being created in the St. John's region out of peat bogs, based on Irish and Swedish experience.

I think I have said enough to suggest that here is a second policy conclusion outside the framework of conventional Maritime thought on the nature of the problem of migration.

7. I might have made some further technical comments on various aspects. (a) The natural increase method attributes all births and deaths to the original population and thus while it may accurately estimate net migration, it underestimates the effects of in-migration. These effects may vary from a region attracting younger immigrants to a region attracting older immigrants. (b) The survival method assumes a constant population. (c) An absorptive capacity model might be very worthwhile developing as an explanation of movement, longer-

developed regions often having a lower capacity than newer regions. (*d*) There is an interesting sociological phenomenon of the earlier-departing females, leaving unbalanced sex ratios with triggering effects for further migration of males. (*e*) I also find it hard to understand serious sex unbalances in migration for non-decision units such as children 0-4, unless there is a greater propensity to move for families with boy children, or else that there is a significant export of male babies from Newfoundland and New Brunswick for adoption elsewhere.

8. I now want to comment on what might be done to make such studies even more useful.

First, I would suggest that a "provincial economic region" framework of analysis is useful for such studies, as a first level of disaggregation, or as a level of re-aggregation. To analyse county data is often meaningless, or at least not the most relevant area of approach. Locational factors can be such that a small area unit can be witnessing economic growth and population decline at the same time. A city may appear to be constant or even declining in population, because its booming suburbs are sprawled across municipal or county lines. One must choose areas of analysis carefully because of the artificiality of county lines, and the large commuting areas we have now.

Just to enable me to organize the data in Tables 16 and 18 in my own mind, within a framework of analysis which enables me to apply other economic and social criteria to this data, I reordered the counties of this region into thirteen provincial economic regions according to the D.D.P. Economic-Administrative Zoning system. Out of these thirteen regions, four had gained, eight had lost, and one had an approximate balance in each period.

Central Newfoundland, western Newfoundland, and Labrador had kept gaining, the Halifax-South Shore region had a balance then a gain, the Annapolis Valley lost then re-established a balance, St. John's-southeast Newfoundland gained and then lost, and the remainder, Prince Edward Island, Cape Breton, northern Nova Scotia, Moncton-southeastern New Brunswick, Saint John-southwest New Brunswick, the upper St. John Valley, and northeastern New Brunswick lost ground consistently. This level of aggregation fades out the local rural-to-urban moves, and emphasizes major shifts out of a region.

The second suggestion is that there are so many explanations, rational and irrational, preferred about the movement of our most important resource, people, that it is about time we really found out something about it. Why do people abandon good land in Nova Scotia? Why do people stay on poor land in Nova Scotia? Why do they go to Toronto instead of Montreal? What happens to these migrants? Why do many become bank presidents or college professors? On the other hand, why do we find that of the Canadian-born chronic drunkenness cases in Toronto's Skid Row, 31 per cent are from the Maritimes and only 2 per cent from western Canada, though the population inflow into Ontario is greater from western Canada?

We have just completed a post-war immigrant survey through D.B.S. which is telling us why people came to and are still residing in Canada, or why they do not intend to reside permanently; if eligible, why they have or have not taken out citizenship. Surely, we should be able to find out not only how many people leave the Maritimes, but also why. Similarly, how many people come into the Maritime areas, and why.

Why has not someone tried to find out? What if the reason for these migrations is the weather, and we find that the counties of low income are also the counties of poor weather, and that the high migration rate is really due to the weather and not the low incomes (which may also be caused by the weather). Or, what if we find in this age of organization man, most of the migration is by posting and not by free choice? Certainly, there is evidence of a considerable proportion of Maritime migration resulting from this source, particularly members of the armed services and their families. The establishment or dissestablishment of a major armed service camp can have major migration effects on a county. If we find such causes, the policy recommendations might be far different from those resulting from other causes.

La Détermination des zones agricoles sous-marginales

GÉRALD FORTIN

DEPUIS QUELQUES ANNÉES, le problème des petites fermes est de plus en plus discuté. L'abandon rapide de l'agriculture, en particulier dans les zones marginales, a forcé les différents gouvernements et les associations agricoles à étudier de plus près les problèmes de ce secteur important de l'agriculture canadienne. Il n'est donc pas superflu d'analyser les critères qui peuvent servir à déterminer le caractère marginal ou commercial d'une ferme ou d'une région agricole. Cette classification des fermes et des régions est une démarche préliminaire nécessaire à toute étude poussée et à l'élaboration de toute politique.

Déjà un certain nombre d'études ont porté sur ce sujet. Ainsi en 1958, la Canadian Agricultural Economics Society consacrait son collogue à ce problème.[1] Les membres de cette association étudièrent tour à tour les moyens de définir une petite ferme, les conséquences de la consolidation, et les politiques les plus appropriées à la situation. De même en 1956, le Bureau fédéral de la statistique offrait une nouvelle définition de la ferme commerciale et des "autres fermes", c'est-à-dire des fermes marginales.

En nous inspirant de ces travaux, nous allons donc essayer de définer le concept de ferme marginale et les critères qui peuvent servir à les repérer. Nous aurons cependant l'occasion de critiquer les conclusions des économistes agricoles et du Bureau fédéral de la statistique et de suggérer certaines modifications qui nous semblent importantes. De plus nous proposerons, dans le cas d'études régionales, l'utilisation de critères simples qui ne sont pas significatifs pour les études à l'échelle nationale mais peuvent le devenir dans le cas d'études plus restreintes. Enfin nous soulèverons le problème du genre de recherches qui doivent précéder l'élaboration de politiques touchant les fermes marginales ou sous-marginales, en proposant la préparation d'un atlas de l'agriculture.

DEFINITION

A leur collogue de 1958, les économistes agricoles ont défini la ferme marginale comme une ferme dont les facteurs de production ne permettent pas des opérations

---

[1] Voir Adélard Tremblay, "Le Problème des fermes marginales", *Agriculture*, janvier, février, 1959, vol. XVI, no. 1.

C.P.S.A. Conference on Statistics, 1961, *Papers.* Printed in the Netherlands.

qui garantissent au fermier des "revenus suffisants pour jouir, dans son milieu, d'un standard de vie convenable pour lui et sa famille." Cette définition est très normative et fait appel à des variables sociologiques plutôt qu'à des variables strictement économiques. Ce standard de vie convenable, il est défini dans *chaque milieu* par un ensemble de valeurs et de normes culturelles. Le "revenu suffisant" variera donc d'un milieu à l'autre, et il sera pratiquement impossible de déterminer un critère de classification unique valable pour tout le pays. Ces variations dans le "revenu suffisant" se manifestent d'ailleurs non seulement d'une province à l'autre mais d'une région à l'autre à l'intérieur d'une même province.

Pour rendre une telle définition réellement opératoire, il faudrait entreprendre des études détaillées des besoins et des aspirations des familles agricoles des diverses régions. De plus, comme le milieu est en voie de transformation rapide sur le plan des valeurs, ces études devraient être entreprises avant chaque recensement. La définition subjective d'un niveau de vie satisfaisant peut varier énormément dans un même milieu au cours de dix ans.

Si toutefois on a besoin d'un critère unique pouvant s'appliquer à l'ensemble du Canada, il faudra fixer arbitrairement un niveau de revenu minimum en bas duquel on jugera qu'un standard de vie minimum est impossible. Il faudra définir quel est le minimum de biens et de services qui doit être assuré à toute famille canadienne. On pourra alors se servir d'une définition strictement personnelle ou d'un niveau de revenu caractéristique de l'ensemble de la population canadienne.

C'est sur ce genre de compromis que semble basée la définition du Bureau fédéral de la statistique. Une ferme est classifiée comme "commerciale" lorsque l'exploitation fournit un revenu brut de $1,200 ou plus, comme "autre", c. à d., "marginale", lorsque le revenu brut est inférieur à $1,200. Le choix du montant de $1,200 est arbitraire. On admet que ce revenu ne permet pas un niveau de vie désirable, mais on affirme que pour atteindre ce revenu l'exploitant doit consacrer suffisamment de temps à l'agriculture pour qu'on puisse considérer son occupation principale comme étant l'agriculture.

Cette définition est donc très différente de celle des économistes agricoles. Le critère de classification n'est plus le standard de vie mais la proportion de son temps ouvrable que le fermier consacre à son exploitation. Même si ce dernier critère du travail non-agricole peut être utile dans certains cas précis que nous décrirons plus bas, son utilisation nous semble assez mal justifiée dans le cas de la définition du Bureau. Tout d'abord on ne nous offre aucune vérification empirique du fait que le cultivateur qui tire un revenu brut de $1,200 de sa ferme y consacre la plus grande partie de son temps. On semble plutôt prendre cette relation comme un postulat, postulat qui est très discutable. Nos quelques observations empiriques dans la Province de Québec indiquent que de tels cultivateurs ne consacrent qu'une partie minime de leur temps à l'agriculture (très souvent la ferme est exploitée par la femme et les enfants plutôt que par le mari). Théoriquement on peut croire qu'un chef de famille qui tire de l'agriculture un revenu de $1,200 (soit un revenu net variant de $500 à $800) doit nécessairement consacrer un temps considérable à une occupation non-agricole afin de s'assurer un revenu minimum pour satisfaire les besoins de sa famille. Théoriquement encore, il y aurait deux exceptions à cette règle: (*a*) le cas de l'individu qui accepte volontairement de vivre dans la misère (plutôt rare); (*b*) le cas de l'individu qui est forcé de vivre dans la misère parce qu'il ne peut trouver aucun emploi non-agricole.

Ce dernier cas peut être assez fréquent dans le nord des provinces des Prairies et dans certaines régions de l'Atlantique. Cettre contrainte du milieu ne rend pas pour autant ces fermiers opérateurs de fermes commerciales.

Reprenant les deux définitions que nous venons d'analyser nous pourrions définir une ferme commerciale comme celle dont les facteurs de production garantissent à l'exploitant un revenu tel qu'il ne doive pas ou peu, pour assurer à sa famille un niveau de vie acceptable dans son milieu, travailler en dehors de la ferme. La ferme marginale serait celle dont les facteurs de production forcent l'exploitant à consacrer à un emploi non-agricole (si un emploi non-agricole est disponible) la plus grande partie de son temps afin d'assurer un standard de vie acceptable à sa famille. Une telle définition nous ramène au problème de la définition d'un niveau de vie acceptable. Des études conduites dans la Province de Québec nous montrent qu'au plan du minimum vital désiré il y a très peu de différences entre le milieu rural et le milieu urbain.[2] Ces indications nous laissent croire que le niveau de vie minimum jugé acceptable même dans les municipalités les plus reculées est beaucoup plus élevé qu'on a pu le croire traditionnellement. A partir de ces données, nous croyons que le revenu minimum qui permettrait de définir une ferme comme commerciale devrait être égal à la moyenne des revenus pour l'ensemble du Canada. Même si l'on tient compte des produits agricoles consommés sur la ferme, ce revenu devrait être un *revenu net* d'au moins $3,000, soit un revenu brut de $4,500 à $5,000.

Le cultivateur qui ne peut s'assurer ce revenu grâce à son exploitation cherche en dehors de la ferme un supplément de revenu ou vit dans la misère.[3]

Jusqu'ici nous nous sommes préoccupés exclusivement de l'adéquation entre le revenu agricole et les besoins de la famille. Un autre aspect important a cependant été négligé. C'est celui de l'utilisation rationnelle des facteurs de production. De deux cultivateurs qui tirent de la ferme un revenu brut de $5,000 et plus, l'un peut le faire grâce à une organisation rationnelle, l'autre, en dépit d'une mauvaise organisation. Ce dernier tire alors le minimum de profit de ses facteurs de production et peut même opérer à perte. A notre avis une ferme ne peut être considérée comme commerciale que si elle est rentable économiquement en plus de garantir un revenu minimum. Tout autre ferme est marginale et est appelée à disparaître à plus ou moins longue échéance. Nous n'avons pas la compétence pour déterminer quel devrait être le taux de rendement minimum du capital investi dans une ferme. Nous croyons cependant qu'une classification socio-économique des fermes devrait tenir compte du rapport entre le profit et le capital investi. Il faudrait se rappeler alors que le revenu net ne constitue pas entièrement un profit vu que l'opérateur doit à même ce revenu se payer un salaire.

Si l'on tient compte de ce nouveau critère on pourrait établir trois catégories principales de ferme. (*a*) les fermes *commerciales* qui assurent un revenu net minimum et qui sont économiquement rentables; (*b*) les fermes *marginales,* qui assurent un revenu net minimum mais qui ne sont pas économiquement viables;[4]

---

[2] "Etude sur les conditions de vie, les besoins, les aspirations des familles salariées de la Province de Québec". Les résultats de l'étude ne sont pas encore publiés. Cette étude est entreprise par le Centre de Recherches Sociales de l'Université Laval.

[3] Voir à ce sujet Gérald Fortin et Marie Tremblay, "Les changements d'occupation dans une paroisse agricole", *Recherches Sociographiques,* vol. 1, no. 4, octobre-décembre, 1960.

[4] La distinction de cette catégorie est très importante au point de vue de l'action agronomi-

et (*c*) les fermes *sous-marginales* dont les facteurs de production ne garantissent pas un revenu net minimum.

Le revenu net minimum pourrait être établi de façon temporaire à $3,000 (revenue brut de $5,000). Ce montant pourrait être revisé à mesure que le niveau de vie canadien augmente. Il est à remarquer d'ailleurs que cette revision d'un recensement à l'autre n'empêcherait pas la comparaison des données des divers recensements. En effet, malgré la variation des montants, les principes de classification resteraient les mêmes, c'est-à-dire l'adéquation entre revenus et besoins et la rentabilité.

Un autre argument qui milite en faveur de l'établissement d'un tel minimum (car à notre sens un revenu brut de $5,000 est un strict minimum) est que non seulement la classe agricole tend à désirer un niveau de vie semblable à celui des classes urbaines mais qu'en justice on doit lui permettre d'atteindre ce niveau de vie.

A ces trois catégories, on pourrait ajouter la catégorie *"ferme des institutions"* déjà définie par le Bureau fédéral de la statistique.

LE CHOIX DES INDICES

Les indices à utiliser pour déterminer à quelle catégorie une ferme appartient sont différents selon que l'analyse se situe au niveau national ou au niveau des régions. Ils vont varier aussi selon que l'on étudie un type de production donné ou tous les types de production à la fois. Enfin les indices qui serviront à mesurer l'adéquation du revenu ne seront pas les mêmes que ceux qui serviront à mesurer la rentabilité des facteurs de production.

1. L'Adéquation du revenu mesurée au plan national

Nous avons déjà mentionné deux indices privilégiés pour mesurer l'adéquation du revenu aux besoins. Le premier est la proportion du temps consacrée à une occupation non-agricole. En utilisant cet indice il n'est pas nécessaire de fixer un montant minimum de revenu. En effet la proportion du temps consacrée au travail non-agricole est en relation directe avec l'adéquation du revenù jugé nécessaire quel que soit le milieu social. L'utilisation universelle de ce critère suppose cependant que tous les cultivateurs qui jugent leur revenu agricole insuffisant peuvent facilement se procurer un emploi non-agricole. Pour l'ensemble du Canada cette hypothèse est fausse. Dans plusieurs régions, les possibilités d'emploi sont à toute fin pratique inexistantes.

Le deuxième critère est le revenu lui-même. Deux problèmes se posent alors: (*a*) la détermination arbitraire d'un seuil (nous en avons déjà discuté), et (*b*) l'estimation de ce revenu net ou brut. Au sujet de ce deuxième problème les solutions

---

que. Les exploitants de cette catégorie ont en effet un capital important engagé dans l'agriculture et consacrent tout leur temps à l'agriculture. A cause de mauvaises méthodes d'administration ou de mauvaises techniques culturales leurs efforts ne sont pas couronnés de succès. Ils ont donc un besoin urgent d'assistance agronomique, beaucoup plus en fait que le fermier commercial qui réussit bien ou le fermier sous-marginal qui déjà abandonne l'agriculture ou qui souvent devrait être encouragé à l'abandonner.

des chercheurs du Bureau fédéral de la statistique sont excellentes. Tout d'abord, on ne peut pas demander directement à un fermier de révéler à un enquêteur son revenu brut, encore moins son revenu net. Ou bien il l'ignore lui-même ou il ne veut pas le révéler. La seule solution est pour le chercheur d'évaluer lui-même ce revenu à partir du recensement de la production et de la valeur moyenne de cette production. Les normes utilisées pour le recensement de 1956 constituent un modèle en ce sens. Il est à remarquer toutefois que cette façon de procéder permet d'évaluer seulement le revenu brut. Le rapport entre revenu net et revenu brut varie considérablement d'un type de production à l'autre, de sorte qu'il est impossible d'établir un rapport constant qui permettrait le passage d'un type de revenu à l'autre.

2. L'Adéquation du revenu mesurée au plan régional

Certains indices, qui étaient inutilisables au plan national, peuvent devenir significatifs lorsque l'univers étudié est une région assez homogène. On pourrait ici multiplier les exemples. Nous ne voulons en signaler que deux.

Lorsque dans une région donnée les emplois non-agricoles sont abondants et facilement accessibles aux cultivateurs, la proportion de temps consacrée à ces emplois est un indice sûr de l'adéquation du revenu agricole Ce serait le cas dans les provinces d'Ontario et de Québec. Le seul problème consiste alors à déterminer un seuil. Quelle proportion de son temps un exploitant peut-il consacrer à une occupation non-agricole avant que sa ferme soit considérée comme sous-marginale? C'est là une question à laquelle on ne peut donner qu'une réponse arbitraire, c'est-à-dire une réponse qui implique un jugement de valeur. La seule façon d'arriver à une réponse objective serait de mesurer sur un échantillon représentatif à la fois le revenu brut (par la méthode décrite en 1) et le travail non-agricole. On pourrait alors calculer le temps moyen consacré à ce travail par les exploitants de ferme dont le revenu brut est supérieur au minimum requis. Cette moyenne pourrait ensuite servir de seuil pour classifier le reste de la population.

Lorsque dans une région on trouve une concentration très forte de cultivateurs spécialisés dans un type de production, des indices secondaires tels que la superficie ou le nombre d'animaux peuvent suffire à classifier les fermes. Par exemple, dans une région de producteurs de grains, la superficie de la ferme est un indice direct du revenu. Un économiste agricole peut alors assez facilement déterminer la superficie minimum que doit avoir une ferme pour cesser d'être sous-marginale.

3. L'Adéquation du revenu selon le type de ferme

Comme nous venons de l'indiquer, pour l'étude d'un univers homogène au point de vue du type de production des indices secondaires peuvent permettre de classifier les fermes. De plus ces indices secondaires (superficie, cheptel, etc.) permettent d'évaluer non seulement le revenu brut mais aussi le revenu net. Il est, en effet, assez facile d'établir un rapport constant entre revenu brut et revenu net à l'intérieur d'une spécialisation donnée. Il serait donc important qu'indépendamment de la classification socio-économique des fermiers on trouve dans le recensement une classification des fermes selon le type de production. Cette dernière classification comprendrait les catégories de *fermes de production mixte,* et de *fermes de production spécialisée*. Cette dernière catégorie pourrait être subdivisée en autant de sous-catégories qu'il serait nécessaire.

En pratique cette classification qui permettrait la classification socio-économique grâce à des indices simples, ne serait pas plus difficile à mettre en oeuvre que le calcul du revenu brut par la méthode du recensement 1956. Par ailleurs elle semble s'imposer à cause de la tendance marquée des fermes canadiennes vers la spécialisation. Il serait important de pouvoir suivre l'évolution historique de ce mouvement vers la spécialisation.

4. La rentabilité des facteurs de production

Pour évaluer la rentabilité des facteurs de production il faut connaître à la fois la valeur des investissements et le revenu net de l'exploitation. Il est toutefois impossible de mesurer ces deux variables directement à cause des réticences des informateurs. On peut par ailleurs connaître assez facilement la nature des investissements et la nature de la production.

Ces derniers indices ne permettent pas d'évaluer la rentabilité des facteurs de production de façon générale. Selon les types de production, la nature des investissements requis de même que la productivité de ces investissements varient énormément. Ce n'est qu'à l'intérieur d'une spécialisation précise qu'une évaluation des rendements est possible à partir seulement d'une connaissance de la nature des investissements et de la quantité de la production.

Ce critère de classification n'est donc utilisable que si une classification selon les types de production existe. Même dans ce cas il reste le problème des *fermes de production mixte* dont le rendement serait difficilement mesurable. Cependant d'après nos observations, la très grande majorité (sinon la totalité) des fermes de production mixte sont des fermes sous-marginales. Si l'on s'en tient à la classification proposée on n'aurait pas à se poser à leur sujet la question de la rentabilité des facteurs de production. Nous proposons en effet que la distinction entre fermes rentables ou non ne soit faite que dans les cas où le revenu agricole est adéquat par rapport aux besoins de la famille. De plus, les quelques fermes de culture mixte qui permettent un revenu adéquat sont les premières à tendre vers la spécialisation. Donc même si un certain nombre de fermes ne pourraient être classifiées selon la rentabilité des facteurs de production parce qu'elles sont des fermes de culture mixte, ce nombre serait assez petit et tendrait à diminuer.

LA DÉTERMINATION DE ZONES

L'intérêt porté aux fermes sous-marginales n'est pas surtout un intérêt scientifique. Ce qu'on vise avant tout c'est la définition de politiques qui pourraient améliorer la situation. Cette position du problème des petites fermes en vue de l'action exige plus qu'une classification et qu'un dénombrement des fermes. Il faut aussi que la *répartition des fermes* selon les diverses catégories soit expliquée. Un facteur important d'explication est la localisation des fermes de divers types. La plupart des autres facteurs sont en corrélation avec la localisation, par exemple, la qualité du sol, le climat, la situation par rapport aux marchés, etc. La marginalité ou la sous-marginalité des fermes ne sont pas des phénomènes qui sont distribués au hasard sur le territoire agricole. Elles se représentent plutôt comme des phénomènes de concentration en zone ou en région. La détermination de ces zones est donc essentielle à l'analyse du problème des petites fermes.

1. Les fermes sont déjà classifiées

Si une classification de toutes les fermes existe, la détermination des zones est assez facile. Il suffit de calculer pour chaque unité de territoire la proportion de fermes appartenant à chaque catégorie. On peut alors porter sur une carte les caractéristiques de chaque unité de territoire. Deux techniques sont possibles. (a) On peut représenter sur une carte la proportion de fermes sous-marginales; sur une deuxième la proportion de fermes marginales, etc. En surimposant ces cartes on a une vue globale de la situation. (b) On peut caractériser univoquement chaque unité selon le type de fermes qui y est le plus fortement représenté. On a alors une seule carte représentant la répartition des territoires selon leur caractéristique principale. Cette carte permet de saisir plus facilement la situation globale.

Dans ce travail de cartographie, l'unité de territoire doit être le plus petit possible. La municipalité ou la subdivision de recensement (selon les provinces) serait la meilleure unité, surtout si on emploie la deuxième façon de représenter graphiquement l'état de l'agriculture. Cette unité est en effet assez petite pour que l'homogénéité soit grande. A cause de cette homogénéité, la classification de l'unité dans la catégorie majoritaire pose assez peu de problèmes; très peu de fermes appartiennent aux autres catégories. Le comté n'offre pas cette homogénéité ou ne l'offre qu'exceptionnellement. Le comté est en effet une unité administrative dont les frontières ne tiennent pas compte des accidents géographiques ou des caractéristiques sociales ou économiques de la population. Tous les types d'agriculture peuvent donc s'y rencontrer dans des proportions à peu près égales. C'est la situation qu'on retrouve dans la Province de Québec par exemple. La classification du comté dans une catégorie particulière devient alors assez arbitraire et peut même fausser l'interprétation de la situation.

2. Les fermes ne sont pas classifiées

Si on ne peut se servir d'une classification de toutes les fermes (par exemple si la classification du Bureau fédéral de la statistique n'est pas jugée satisfaisante pour les fins qu'on poursuit) la détermination de zones n'exige pas nécessairement qu'on entreprenne un recensement exhaustif de toutes les fermes de la région qu'on veut étudier.

Sans doute on pourrait procéder à un recensement par échantillon, mais il existe des techniques plus rapides et moins coûteuses qui semblent aussi efficaces et aussi sûres. Une de ces techniques que nous avons utilisée, sur un territoire comprenant six comtés, est l'évaluation objective des bâtiments de ferme.[5] En parcourant en automobile tous les rangs des municipalités étudiées, nous avons classifié les fermes en trois groupes: (a) les fermes où à la fois la maison et les bâtiments sont couverts de peinture et bien entretenus (fermes commerciales); (b) celles où seuls les bâti-

---

[5] Gérald Fortin et Emile Gosselin, "La professionalisation du travail en forêt", Recherches Sociographiques, vol. 1, no. 1, janvier-mars, 1960.

ments sont bien entretenus (fermes marginales); (c) celles où seule la maison est bien entrenue et celles où ni la maison ni les bâtiments ne sont en bon état (fermes sous-marginales). Cette classification sommaire s'est avérée exacte dans 90 pour cent des cas lorsque nous l'avons comparée à une classification basée sur un questionnaire et des interviews dans trois municipalités.

Il ne s'agit pas là d'un critère universel. C'est à partir d'une certaine connaissance du système de valeurs et de la mentalité des cultivateurs de la région étudiée que nous avons pu choisir ces indices. Dans d'autres régions, on pourrait cependant trouver d'autres indices qui rempliraient le même rôle.

Une technique qui est utilisable dans n'importe quelle région est le recours à des informateurs-clés. Dans toutes les provinces, il existe des agronomes de comtés, des inspecteurs gouvernementaux, des gérants de coopérative, des agents commerciaux, les leaders religieux ou civils qui connaissent intimement un petit secteur du milieu agricole. En recueillant ces connaissances diverses et en confrontant les renseignements recueillis afin d'éliminer les biais possibles des informateurs, on peut reconstituer une classification des municipaltés agricoles qui soit très valable.

C'est ainsi que dans un mois nous avons pu sans quitter la ville de Québec classifier toutes les municipalités agricoles de la Province de Québec. Profitant de congrès d'agronomes de comté, de la présence d'inspecteurs gouvernementaux et d'autres experts, nous avons recueilli suffisamment de données sur chaque municipalité pour les classifier selon l'adéquation du revenu agricole aux besoins des familles.[6] Comme nous l'avons signalé plus haut dans la Province de Québec il existe suffisamment de possibilités d'emplois non-agricoles pour que la proportion du temps que le cultivateur consacre à ces emplois soit un indice sûr de l'adéquation du revenu agricole aux besoins. C'est donc ce critère dont nous nous sommes servis pour évaluer les municipalités. Aux informateurs-clés nous avons demandé d'évaluer, pour chaque municipalité qu'ils connaissaient très bien, la proportion de cultivateurs qui avaient des emplois non-agricoles de même que la proportion du temps consacrée à ces emplois. Malgré une légère tendance de tous nos informateurs à surestimer la proportion du temps consacrée à l'agriculture, notre classification s'est avérée juste dans tous les cas où nous avons pu la vérifier par des techniques plus rigoureuses. Dans d'autres régions, le critère du travail non-agricole ne pourrait être employé, mais d'autres critères pourraient être définis selon la situation.

UN ATLAS DE L'AGRICULTURE CANADIENNE

La détermination des zones agricoles définies d'après leur catégorie économique n'est qu'un premier pas vers une analyse réaliste de l'agriculture canadienne et vers la définition de politiques efficaces. Une analyse tant soit peu complète suppose que les relations qui existent entre la qualité du sol, le climat, les types de production, les marchés, les rendements, la valeur des investissements, etc. soient établies, non seulement au plan général et abstrait, mais au niveau de chaque région.

L'agriculture canadienne est en effet, tellement diversifiée qu'une explication et une planification à l'échelle nationale sont illusoires sauf au plan des principes

[6] Gérald Fortin, "Une classification socio-économique des municipalités agricoles du Québec", *Recherches Sociographiques*, vol. 1, no. 2, avril-juin, 1960.

de base. Toute planification en ce domaine doit tenir compte des caractéristiques et des problèmes régionaux.

A cette fin, le représentation cartographique des principaux aspects de l'agriculture est un outil préalable indispensable. Déjà un bon nombre de recherches ont été faites dont les résultats se traduisent graphiquement (surveys pédologiques), ou pourraient être facilement traduits (classifications du recensement). En fait la plupart des données du recensement agricole pourraient être représentées sur des cartes dont l'unité territoriale serait la subdivision de recensement.

A partir de ces données, on pourrait constituer un atlas de l'agriculture canadienne qui, entre autres, pourrait couvrir les aspects suivants: qualité du sol, climat, marchés, superficie des fermes, capitalisation, type de machinerie, type de production, classification économique, niveau de vie des familles, augmentation (ou diminution) de la population agricole et non-agricole, augmentation (ou diminution) du nombre de fermes, mouvement vers la spécialisation ou la concentration, etc.

Cet atlas est sans doute irréalisable dans sa forme idéale dès maintenant. Trop de recherches restent à faire. Nous croyons cependant qu'il serait utile d'en préparer immédiatement une version préliminaire. Même incomplet un tel instrument de travail constituerait une somme et un résumé de toutes les recherches effectuées à date. Ce serait déjà là une fonction très importante. Plus important encore, il permettrait de déterminer avec précision les vides dans notre connaissance du problème. Ainsi une planification de la recherche deviendrait possible ainsi que la mise en oeuvre d'une coopération plus étroite entre les chercheurs de diverses régions et de diverses disciplines.

Discussion

N. KEYFITZ and J. HENRIPIN

LA CONTRIBUTION du professeur Fortin nous rappelle que, pour compter quoi que ce soit, il faut le définir, et que les définitions qui suffisent pour la vie courante ne suffisent point pour la statistique. Elles doivent être plus simples dans l'application; nous ne disons pas plus précises, parce que la précision dépend de l'objectif. Mais il faut que tout le monde soit d'accord sur le fait que, d'après la définition du recensement, A est une ferme, et B n'en est pas une. Cette simplicité, cette objectivité, ne s'obtiennent pas sans laisser de côté des faits et des circonstances importants. Et surtout on doit laisser de côté toute considération de valeur. Le statisticien officiel est situé entre mille pressions sociales et politiques; ce qu'il fait est objectif en ce sens que son travail doit servir à tout le monde; si vous exigez qu'il façonne sa définition exactement selon vos besoins, il est possible que votre voisin soit déçu, ou même que le statisticien ne puisse faire son travail du tout.

Le Bureau fédéral de la statistique, après avoir reçu beaucoup de conseils de l'extérieur, a décidé de donner des détails pour les fermes dont la valeur brute de la production est de $1,200 et plus. Ces fermes sont appelées "commerciales" pour éviter une terminologie trop compliquée. En fait, il ne faut voir dans ce terme rien de normatif; pour éviter toute confusion, il faut lire, au lieu du mot "commerciale", l'expression: "groupe de fermes pour lesquelles le Bureau a l'intention de donner des renseignements détaillés". Avec cela, on peut utiliser, suivant ses besoins et ses critères, une limite de $1,200, $2,500 ou $3,750, etc. La classification comprend tous ces chiffres, jusqu'à $25,000. Si le professeur Fortin choisit de couper à $5,000, le vendeur de tracteurs lourds peut préférer couper à $15,000. C'est ainsi que les statistiques officielles sont mises sur le marché, le marché des affaires ainsi que le marché des investigations scientifiques.

Il est impossible pour le Bureau fédéral de la statistique, dans le présent ou dans le futur, d'adopter quelque critère que ce soit comme "minimum de biens et de services qui doit être assuré à toute famille canadienne." Si le Bureau faisait des spéculations sur de telles choses, il négligerait sa vraie mission, qui est de donner au public des chiffres aussi exacts que possible. Mais il est heureux que les professeurs de Laval s'intéressent à des problèmes aussi importants. Ils cherchent à trouver non seulement le niveau de vie du cultivateur, mais aussi ses besoins et ses aspirations qui ne sont pas satisfaits. Ainsi, dans leurs études, ils ont trouvé qu' "au plan du minimum désiré, il y a très peu de différence entre le milieu rural et le milieu urbain." Il n'y a aucun doute que les besoins s'accroissent partout dans le monde, même dans des régions beaucoup plus isolées que les campagnes canadiennes, et nous serons bien heureux de voir en détail les données de M. Fortin et ses collègues montrant comment nos cultivateurs se modernisent au point de vue de la psychologie des besoins.

Mais au-delà de toute différence entre le professeur Fortin et le Bureau fédéral de la statistique, nous trouvons des difficultés d'ordre économique dans les suggestions de ce papier. Nous posons simplement une question et nous ne serons pas assez temeraires pour essayer d'y répondre. Le professeur Fortin considère qu'une ferme où le cultivateur travaille dans d'autres emplois doit être considérée comme sousmarginale; est-ce qu'il veut dire que cette ferme doit disparaître? N'est-il pas possible qu'une ferme dont le propriétaire travaille une partie de l'année dans les bois ou dans une usine soit plus durable, et qu'en même temps le propriétaire jouisse d'un niveau de vie beaucoup plus satisfaisant que s'il vivait sur une ferme semblable sans avoir la possibilité de faire du travail supplémentaire?

Some Calculations
Relating to Trends and Fluctuations
in the Post-War
Canadian Labour Market

FRANK T. DENTON

THIS PAPER IS CONCERNED with the Canadian labour market in the period 1947-60. More specifically, it is concerned with trends and variations in the labour force, in unemployment, and in employment in the aggregate and in different industrial divisions of the economy. Its aim is to throw some light on questions such as the following. Which industries have contributed most to the growth of total employment throughout the period and which industries have contributed least (or have made a negative contribution)? Which industries are the most volatile sources of employment in the short run and which are the most stable? To what extent have short-run changes in the size of the labour force contributed to fluctuations of unemployment and to what extent have these fluctuations been related to fluctuations of employment? In what degree do employment fluctuations in particular industries conform to fluctuations of aggregate employment and in what degree do they occur independently of such fluctuations? Which industries, by virtue of their size, their volatility, and their conformity with the behaviour of aggregate employment, have contributed most to the short-run movements of aggregate employment and unemployment? What differences of behaviour are exhibited by different series of employment statistics and, in so far as there are differences, what inferences can be drawn as to actual occurrences in the labour market and as to the relative usefulness of the different statistical series as measures of change?

It was decided at the outset to restrict the analysis to annual data, and consequently no attention is given to seasonality, important as this is in the Canadian labour market. It is recognized, of course, that annual averages, while they can be expected to reflect longer-run movements quite faithfully, do not reflect with complete adequacy all of the characteristics of shorter-run non-seasonal behaviour which one might wish to take into account. In particular, it is impossible by the examination of such averages to determine precisely the cyclical turning points in employment and unemployment or to ascertain the various lead or lag properties of different series. However, it was felt that calculations based on annual data would be quite satisfactory, and in retrospect it is considered that any problems

C.P.S.A. Conference on Statistics, 1961, *Papers*. Printed in the Netherlands.

arising from this source are far outweighed by problems of other kinds which are noted throughout the paper. Moreover, the use of annual rather than seasonally-adjusted intra-annual data greatly reduced the burden of computation, and this was an important consideration.

The analysis makes use of simple statistical techniques. In particular, it involves the calculation of least-squares trend lines, the examination of the properties of these trend lines, the calculation of the standard deviations of the variations around the trend lines, and the correlation of these variations with the variations of unemployment and of different employment measures. Its chief merit—if it has any merit—lies in its attempt to treat systematically a reasonably large body of industrial data, to examine employment statistics obtained from different sources and by different methods rather than to rely on any single set of statistics, to distinguish regorously between shorter-run and longer-run movements, and to decompose the changes in total employment and unemployment in a manner which permits a theoretically exact quantitative assessment of the contributions of movements in different industries to movements of the aggregates.

The framework within which the analysis has been conducted is described in a formal manner in Section I, and the statistics which have been analyzed are discussed in Section II. The actual numerical results of the calculations are presented in a number of tables at the end of the paper, and some notes pertaining to these results are provided in Section III.

Finally, it may be noted that the statistical techniques used in this paper have wide applicability. They can be used to analyse systematically trends and fluctuations in any situation in which a time series of totals can be expressed as the sum of the time series of its components or, more generally, in any situation in which the time series for some variable can be expressed as a linear function of the time series for a set of other variables.

I. THE ANALYTICAL FRAMEWORK

A formal description of the analytical framework proceeds as follows. For any variable X (employment, unemployment, labour force), it is assumed that at any given time there is a trend component $\bar{X}$ and a residual or deviation from trend x, so that

$$X = \bar{X} + x.$$

Furthermore, the conditions are imposed that if X is equal to the sum of a set of variables X_i ($i = 1, 2, \ldots, n$) then $\bar{X}$ is equal to the sum of the trend components of the variables and x is equal to the sum of the residual components. (These are not independent conditions, of course.)

$$X = \Sigma X_i$$
$$\bar{X} = \Sigma \bar{X}_i$$
$$x = \Sigma x_i$$

Trend values are given by equations of the form

$$\bar{X} = b_0 + b_1 t + b_2 t^2$$

and it follows that

$$b_0 = \Sigma\, b_{0t}$$
$$b_1 = \Sigma\, b_{1t}$$
$$b_2 = \Sigma\, b_{2t}$$

where the summations are over the trend equations[1] for all of the X_t. The trend equations are fitted to the data by the method of least squares and it is a convenient property of least-squares regressions that they satisfy the conditions just stated: the least-squares values of the coefficients of the trend equation for X are equal to the sums of the least-squares values of the coefficients of the trend equations for the X_t.[2]

As a matter of convenience, t is set equal to zero at the midpoint of the period (in the present case, the period 1947-60 with midpoint at the end of the year 1953). This, of course, makes the summation of t over the whole of the period equal to zero. Also, a scale is selected such that a unit change in t is equal to one year.

In analysing changes over time, both longer-run and shorter-run movements are of interest and it is important not to confuse the two. Longer-run movements will be defined as *movements along the trend line* (that is, changes in $\bar{X}$), and shorter-run movements as *variations in the deviations from the trend line* (that is, changes in x). Thus, if X' and X'' are the values of X at two points of time, the difference may be broken into its longer-run and shorter-run components:

$$X'' - X' = (\bar{X}'' - \bar{X}') + (x'' - x').$$

In discussing the different types of movement, reference will be made to "trend growth" (growth along the trend line), "trend acceleration" (acceleration along the trend line), and "residual fluctuation" (changes in the deviations or residuals).

The rates of trend growth and trend acceleration are given by the first and second derivatives of the trend function with respect to time, and are equal, respectively, to $b_1 + 2b_2 t$, and $2b_2$. Let G represent the annual rate of trend growth averaged over the period (or, alternatively, the annual rate at the middle of the period, since this is the same), and let A represent the annual rate of trend acceleration (which is constant). It is immediately evident that $G = b_1$ and $A = 2b_2$. As before, $G = \Sigma\, G_t$ and $A = \Sigma\, A_t$, and this allows a simple determination of the contribution of each, of the X_t to the trend growth and trend acceleration of X. Furthermore, $g = G/X_m$ and $a = A/X_m$ are taken to be suitable relative measures, where X_m is the mean value of X (and of $\bar{X}$) over the period. It may be noted, in particular, that g is preferred to the usual

---

[1] Relationships of this kind hold for equations of any degree in t. However, quadratic equations were considered most appropriate for the present study. Linear equations would not allow for changes in (arithmetic) growth rates. On the other hand, cubic or higher-degree equations might reflect short-run (cyclical) movement as well as trends, and this would be most undesirable since the aim of the analysis is to separate these two types of movements,

[2] As a general demonstration, suppose that $y = \Sigma\, y_t$ expresses some scalar variable as the sum of its parts, and consider the regressions of y and y_t on a vector of "independent" variables $z = [z_1\ z_2\ \ldots\ z_p]$. The least-squares values of the vectors of coefficients are given by $k = M_{zz}^{-1} M_{zy}$ and $k_t = M_{zz}^{-1} M_{zy_t}$ where M denotes a matrix of second moments and the subscripts indicate the variables from which the moments are calculated. Summing over all components of y, $\Sigma\, k_t = \Sigma\, M_{zz}^{-1} M_{zy} = M_{zz}^{-1} \Sigma\, M_{zy_t}$ But $\Sigma\, M_{zy_t} = M_{zy}$, and hence $\Sigma\, k_j = k$.

compound growth rate calculated from the actual values of X at the beginning and end of the period since the latter is unduly affected by the situations obtaining at the terminal dates and may reflect both longer-run and shorter-run change. For example if the series is cyclically volatile, and if it is at a cyclical peak at the beginning of the period and a cyclical trough at the end, the ordinary compound growth rate calculation may be deceptively low; in the converse situation it may be deceptively high. In the calculation of g the terminal values also affect the result but their effect is no larger than that of any of the intermediate values and so the danger of obtaining a misleading result is not nearly as great.

In analysing short-run fluctuation, the summary measures which are employed are the standard deviation of the residuals about their mean (which is zero), $S(x)$, and the ratio of this standard deviation to the mean value of X over the period, $\bar{s}(x) = S(x)/X_m$. It is desirable also to have some means of evaluating the contribution of residual fluctuation in each of the components to residual fluctuation in the total. The decomposition of $S(x)$ is not quite as straightforward as the decomposition of G and A, for the standard deviation of a total in general is not equal to the simple sum of the standard deviations of its parts. However, it is easily shown that the standard deviation of a total is equal to the *weighted* sum of the standard deviations of its parts where the weights are the coefficients of correlation of the parts with the total. Thus,

$$S(x) = \Sigma\, r_i\, S(x_i)$$

where r_i is the coefficient of correlation of x and x_i.[3] The product $r_i\, S(x_i)$, which is equivalent to the covariance of x and x_i divided by the standard deviation of x, provides a simple and intuitively reasonable measure of the contribution of the fluctuation of x_i to the fluctuation of x. Each of the parts is regarded as contributing to fluctuation of the total some fraction (positive or negative) of its own fluctuation, the fraction depending on the degree to which its movements support or offset the movements of the total. A component which moves in close accord with the total (correlation coefficient close to unity) contributes most of its fluctuation, a component which moves quite independently of the total (correlation coefficient close to zero) contributes very little, and a component which moves in an opposing manner (correlation coefficient less than zero) makes a negative contribution. Finally, the correlation coefficient is of interest in itself as a measure of "conformity," that is, as a measure of the degree to which the behaviour of x_i conforms or does not conform to the behaviour of x. In particular, it is of interest to determine whether or not a given series exhibits cyclical behaviour similar to that which is exhibited by the total.

The argument of this section has proceeded from the assumption that X is equal to a simple sum of the X_i. However, it can easily be extended to the more general case of a linear function. Suppose $X = \Sigma\, w_i X_i$, where the w_i are constants. Set $Y_i = w_i X_i$, so that $X = \Sigma\, Y_i$. One could then proceed exactly as before, fitting regression equations to the Y_i instead of the X_i. But the only effect of the trans-

---

[3] The coefficient of (linear) correlation of x and x_i is given by the ratio of the covariance to the product of the standard deviations, $S(x,x_i) / S(x)\, S(x_i)$. Multiplying by $S(x_i)$ and summing over i gives $\Sigma\, S(x,x_i) / S(x)$, and since $\Sigma\, S(x,x_i) = S(x,x)$ or $S(x)^2$ this is equal to $S(x)$.

formation on the coefficients and the standard deviation of the residuals would be that of multiplying them all by w_i. Making a minor and obvious change in the subscript notation,

$$G_{Y_i} = w_i\, G_{X_i}$$
$$A_{Y_i} = w_i\, A_{X_i}$$
$$S(y_i) = w_i\, S(x_i)$$

and the coefficients of correlation are unchanged. Reverting to the original subscript notation, it follows that if $X = \Sigma\, w_i X_i$, then

$$G = \Sigma\, w_i\, G_i$$
$$A = \Sigma\, w_i\, A_i$$
$$S(x) = \Sigma\, w_i/r_i S(x_i)$$

If the linear function contains an additive constant one may regard this as the coefficient of a "dummy" variable which is always equal to unity.

The foregoing represents a formal and quite general description of the analytical framework and it must now be related more specifically to the problems at hand. The relevant variables are the labour force (F), unemployment (U), and employment in each of n industries (E_i), and the following identities provide a starting point:

$$U = \bar{U} + u$$
$$F = \bar{F} + f$$
$$E_i = \bar{E}_i + e_i$$
$$U = F - E_1 - E_2 - \ldots - E_n$$
$$\bar{U} = \bar{F} - \bar{E}_1 - \bar{E}_2 - \ldots - \bar{E}_n$$
$$u = f - e_1 - e_2 - \ldots - e_n$$

By performing the appropriate calculations it is possible to express the rate of trend growth of unemployment in terms of the rate of trend growth of the labour force and the rate of trend growth of employment in each industry. Similarly, it is possible to express the rate of trend acceleration of unemployment and the standard deviation of the unemployment residuals in terms of labour force and industrial employment components. It is also possible to analyse changes in total employment in the same way, expressing them in terms of growth, acceleration, and fluctuation in the various industries which make up the total. The results of such calculations, and of the other related calculations discussed above, are presented in tables at the end of this paper and they are considered below. First, however, it is necessary to examine the nature and sources of the statistics on which the calculations are based.

II. THE STATISTICS

The analysis required annual averages of the labour force, unemployment, and employment, the latter broken down in some detail by industry, for the period 1947-60. In the case of the labour force, unemployment, and aggregate employment, the sample estimates provided by the Dominion Bureau of Statistics' labour force survey[4] were accepted uncritically, subject only to the qualification that the absence of Newfoundland from the survey prior to October, 1949, necessitated some upward adjustments to the estimates for the early years. It should be noted also that before November, 1952, the survey, which is now a monthly survey, was conducted at roughly quarterly intervals and that therefore the averages for the period 1947-52 are based on only four observations in each year while the averages for later years are based on twelve observations. However, it appears that this has introduced no significant discontinuities into the series, although the averages for the earlier years may be subject to somewhat greater sampling variability.

In the case of the industrial components of employment, the situation is not as simple. There are several sets of employment data, obtained from different sources and by different methods, and it is a fact well known to persons familiar with the various sets that they do not always appear to behave in a consistent manner, even after appropriate allowance is made for differences of concept and coverage. Rather than make an arbitrary selection it was decided to examine all of the major sets of data, and to this end four separate sets of calculations were carried out. These will be referred to as calculations A, B, C, and D.

The A calculations are based on annual averages of the labour force survey sample estimates of employment in each of eleven industrial divisions.[5] As in the case of the other series obtained from this source, adjustments were made to compensate for the exclusion of Newfoundland in the early years, and for years prior to 1953 each average is based on only four observations. On the one hand, these are the most comprehensive estimates available in the sense that they are available for every major industrial group and that they cover all types of workers—paid workers, own-account workers, employers, and unpaid family workers, to use the conventional terms. On the other hand, the estimates are subject to sampling variability and even a casual inspection suggests that in certain industries this may have been a factor of some importance (estimates of sampling variances for the annual averages in the various industries are not available, and for this reason it is impossible to be much more precise).

The B calculations are also based on employment estimates obtained from the labour force survey, but in this case on the estimates of paid workers only, that is, workers other than self-employed persons and unpaid workers in family enterprises. Again the estimates were adjusted for the exclusion of Newfoundland in the early years, and again the average for each year prior to 1953 is based on only four observations. These estimates are also subject to sampling variability, of course.

The C calculations are based on a special set of employment estimates constructed in the following manner. The annual average employment indexes derived from

---

[4] See the D.B.S. monthly publication no. 71-001, *The Labour Force*. It may be noted that some changes of terminology and definition have recently been made in this survey. In particular, the terms "employed" and "unemployed" have been introduced explicitly, and persons on temporary layoff with instructions to return to work within thirty days are now regarded as "unemployed", whereas formerly they were regarded as "persons with jobs." For background material and the recommendations on which these changes were based, see *Report of the Committee on Unemployment Statistics* (Ottawa, 1960). It is to be understood that all references in this paper to labour force survey estimates are to the estimates on the new basis.

[5] These, and the annual average estimates of paid workers used in the B calculations, are contained in special tables available on request from D.B.S.

the D.B.S. monthly survey of establishments with fifteen or more persons on pay-roll[6] were used for each industry to project from the 1951 census figures for wage-earners, the latter adjusted to exclude persons without jobs and to include employed wage-earners who did not report their industrial attachments.[7] The projections were carried out in as much industrial detail as the published indexes would permit and the results were then aggregated to obtain totals for the industrial groups which were to be included in the analysis. This resulted in a set of absolute employment measures (as opposed to a set of indexes). It also resulted, in many cases, in a reweighting of the establishment survey indexes according to the census distribution of employment, and presumably this is a desirable feature since the published indexes relate only to larger establishments and the implicit weights on which they are based are therefore determined in part by differences in the size "mix" of establishments in different industries, as well as by the over-all size of each industry in relation to other industries. This set of estimates provides much more industrial detail than any of the other sets (indeed, because of limitations of time it was necessary to ignore, with some reluctance, a large part of the available detail, particularly in the manufacturing industries). However, the estimates do not cover self-employed persons or unpaid workers in family enterprises. Moreover, while they are not subject to random sampling variability, as are the labour force survey estimates, they may be subject to a type of bias, for in so far as the smaller establishments in any industry differ from the larger establishments with respect to employment trends or variations, the indexes on which the estimates are based will fail to represent accurately the movements of employment in the industry as a whole. (The extent to which this is true depends also, of course, on the distribution of total employment between smaller and larger establishments in the industry). The data on which the B and C calculations are based are sufficiently similar with respect to definition of employment to permit rough comparisons of the results obtained for each industrial group and it is of interest to make such comparisons. That there are some significant differences is evident from the tables at the end of the paper, and these differences are discussed below.

Finally, the D calculations are based on the employment figures provided by the annual D.B.S. census of manufacturing.[8] Adjustments were made to the averages for the early years to compensate for the exclusion of Newfoundland. Also, because 1959 and 1960 data were not yet available, estimates had to be made for these years, and these were made by projecting from 1958 on the basis of

---

[6] See the D.B.S. monthly publication no. 72-002, *Employment and Payrolls* and no. 72-201, *Annual Review of Employment and Payrolls*. These contain descriptions of methods, and the monthly publication provides measures of survey coverage in different industrial divisions.

[7] The adjustments were of necessity somewhat arbitrary, involving the simple allocation of the "not stated" groups on the basis of the distribution of persons who did report their industries; as the magnitudes of the adjustments were extremely small, though, this is of very little importance. In addition, it may be noted that in some cases there are residual groups for which no indexes are available. For example, indexes are available for food and beverage manufacturing and for all of its major components (meat products, dairy products, etc.) but there is a small group within food and beverage manufacturing for which no index is available. A little experimentation suggested that the best way of treating such a group was to assume that its index behaved in the same way as the published index for the industry as a whole (e.g., the published index for the food and beverage industry).

[8] See the annual D.B.S. publication no. 31-201, *General Review of the Manufacturing Industries of Canada*.

the percentage changes in the specially constructed C series discussed in the previous paragraph. The census of manufacturing, unlike the monthly employment survey, attempts to cover all manufacturing establishments, regardless of size. However, as it is otherwise quite similar to the monthly survey, both in definition of employment and in method of measurement, it is of interest to compare the results of B, C, and D calculations for total manufacturing, and the results of C and D calculations for the major groups within manufacturing. Again, the differences are discussed below.

III. THE RESULTS

The numerical results are presented in Tables 1 to 13. Because of limitations of time it has not been possible to extract and reflect adequately on all of the points arising out of the calculations and this section merely provides some rough notes relating to the more obvious points. These notes can hardly be regarded as an adequate interpretation of the results, and they are presented merely as a guide for the reader who may wish to work his way through the tables. As a matter of convenience, the notes are grouped under three headings: trend growth, trend acceleration, and residual fluctuation.

1. Trend Growth

The tables of interest under this heading are Tables 2 and 3. Table 2 presents the calculated values of G, the annual average arithmetic rate of trend growth, and Table 3 presents the corresponding values of g, the ratio of G to the average 1947-60 group size (the average group size figures appear in Table 1). In these and all other tables, four sets of calculations are presented, based on the four sets of employment data discussed in the previous section. The figures in the last row of each table represent the calculations for "employment not covered," i.e., the difference between the D.B.S. labour force survey estimate of total employment (which has been accepted uncritically) and the sum of the employment figures available from the less comprehensive sets of data used in the B, C, and D calculations. Because the "employment not covered" group varies greatly in content between the different sets of data, the figures in this row are not comparable with each other. Also, to the extent that there is inconsistency between the A-B estimates and the C and D estimates, the residual "employment not covered" figures in the C and D columns will be subject to error.

The picture of growth in the labour market is clear in outline, if somewhat blurred in detail. The disparity between the rates of trend growth of the labour force and total employment, and the concomitant rise of unemployment may be noted. Farm employment has declined very sharply throughout the period, and the calculated annual trend growth of total employment of 82,700 is the result of an annual increase in non-agricultural employment of 119,300 offset in part by an annual decline of 36,600 in agriculture. Apparently there has also been a decline in fishing and trapping. (It must be noted, though, that the numbers of people involved are relatively small and the A and B estimates of employment are undoubtedly subject to considerable sampling variability which may affect both longer- and shorter-run movements in these series). In all of the other major industrial groups, employment has risen, but it is immediately evident that it has risen much more rapidly in the so-called service-producing industries (all industries from transportation down) than in the goods-producing industries. In communication, public utilities, retail and wholesale trade, finance, insurance, and real estate, and service[9] the values of g are all in excess of 3 per cent. The only exception within the service-producing industries occurs in transportation where there have been some notable shifts between components—rapid growth in air transport and truck transportation, and declines in the railways, water transportation, and urban and interurban transportation. The net result of these shifts has been that employment in the transportation group as a whole has increased much less rapidly than employment in other service-producing industries.

Among the goods-producing industries, employment has risen more rapidly in construction and mining than in manufacturing. The trend growth in the mining group is the net result of internal shifts—very fast growth in the extraction of oil and natural gas and in the mining of metals other than gold, and sharp declines in coal- and gold-mining. Within manufacturing, trend growth has been more rapid in durable goods industries than in non-durable goods industries, each group take as a whole, and among the non-durable industries there are several cases of decline.

The various sets of calculations displayed in Tables 2 and 3, taken separately and together, provide sufficient evidence to warrant the foregoing remarks. However, more detailed analysis is hampered in some cases by problems of apparent inconsistency. Running down the tables, the first example occurs in forestry where the B and C values are seen to have opposite signs. On the basis of the evidence it is not clear whether employment in forestry should be regarded as having risen or fallen in the long run. (It should be noted, however, that this industry is notoriously volatile, with the result that the short-run movements tend to swamp the longer-run movements, and that it is also a notoriously difficult industry for which to obtain accurate measurements.)

In mining, the agreement between the B and C calculations is very good, but in manufacturing there is a substantial difference, with the B calculations yielding an arithmetic trend growth rate which is 1.5 times the C rate. In construction there is also a substantial difference, the B rate in this case being about 1.7 times the C rate.

The first thought that cames to mind in seeking an explanation for the difference in manufacturing is that the indexes used in obtaining the C estimates of employment relate only to establishments with fifteen or more employees, and that possibly growth has been substantially more rapid in smaller establishments. (This seems unlikely, but it is nevertheless a possibility to be considered.) Apparently this is not so, for the C calculations generally agree quite well with the D calculations which are based on D.B.S. census of manufacturing data relating to establishments of all sizes. There are some differences in particular industries, but on the whole, and allowing for minor variations in classification procedures, the agreement is good. At the level of total manufacturing the C and D results are almost iden-

---

[9] "Service" is to be distinguished from the much broader "service-producing" group; "service" includes only government, community service, business service, recreational service, and personal service.

tical. In view of this, the explanation of the difference between the B and C results should probably be sought elsewhere. However, no further attempt will be made here, the present purpose being merely to draw attention to the difference.

The agreement between the B and C calculations is considerably better in the service-producing industries. In the transportation, storage, and communication group the results are very close. Elsewhere there are large differences, most notably in trade, but on the whole the differences provide much less cause for concern than the differences in manufacturing, construction, and forestry.(Note, however, that because the C data exclude government, community service, and domestic service, no meaningful comparison is possible for the large service group.)

In concluding these remarks on trend growth, it must be observed that there is a pronounced tendency for the B calculations to yield higher rates than the C calculations. In terms of the arithmetic rates of Table 2, the B rates are higher in six of the eight groups for which meaningful comparison is possible, and in terms of the relative measures of Table 3, the B calculations are higher in seven of the eight groups. In other words, there has been a pronounced tendency for the estimates of paid workers provided by the labour force survey to rise more rapidly in the period 1947-60 than the corresponding indexes of employment provided by the monthly survey of establishments, even after the latter indexes are reweighted to achieve closer conformity with the actual industrial distribution of employment.

2. Trend Acceleration

There has been a noticeable slowing down of the rate of economic growth in Canada. An examination of the last three post-war cycles of output, investment, and other series, reveals that the expansions have become weaker and conspicuously shorter in duration.[10] In view of this, it is appropriate to ask whether there has been a similar slowing down of growth in the labour market.

The tables of interest are Tables 4 to 6. Table 4 presents the calculated values of A, the annual rate of trend acceleration, and Table 5 presents the corresponding values of a, the ratio of A to the average 1947-60 group size. Table 6 presents the increases or decreases in the percentage rates of trend growth from the beginning of the period to the end of the period. The rate at the beginning of the period is taken to be the difference between the 1947 and 1948 trend values divided by the average of these values, and the rate at the end of the period is taken to be the difference between the 1959 and 1960 trend values divided by their average. The 1947-48 annual trend rate based on A calculations for service, for example, was 0.62 per cent and the 1959-60 rate was 6.09 per cent. The increase in Table 6 of 5.47 percentage points is the difference between 6.09 and 0.62. It may be noted that an increase may arise also from a decrease in rate of decline. For example, the A estimates for agriculture exhibited a declining trend throughout the 1947-60 period but the rate of decline along the trend line was less rapid at the end than at the beginning, and the 1.07 percentage point increase in Table 6 is the difference between the 1947-48 rate of —4.47 per cent and the 1959-60 rate of —3.40 per cent.

---

[10] This has been documented by Professor Wm. C. Hood in his report to the Special Committee of the Senate on Manpower and Employment. See no. 2 of the *Proceedings* of the committee (Ottawa, 1960).

The evidence of Tables 4 and 5 indicates that the labour force, total employment, and unemployment have all experienced acceleration of growth rates. It is apparent, though, that the service industry has accounted for a remarkably large part of the acceleration in the case of total employment. The slowing down of the rate of decline in agriculture has also contributed substantially, though not nearly as much as the acceleration in service. Among non-agricultural industries other than service, there has been widespread deceleration. However, there is again some notable disagreement between the different sets of calculations. In manufacturing, the *C* calculations agree quite well with the *D* calculations but disagree sharply with the *A* and *B* calculations, and there is substantial disagreement in other industries as well, particularly in the transportation, storage, and communication group.

It is possible, of course, to have an increase in the arithmetic rate of growth but a decrease in the percentage rate, and the changes in percentage rates recorded in Table 6 indicate more clearly than the other tables the widespread tendency for the rate of growth of employment to fall outside of service and agriculture. Apart from these two industries, there are very few figures in the table that do not have minus signs. (The different results obtained for manufacturing and for some other industries are still cause for concern, though.)

If one can believe both the *A-B* and the *C* calculations, the remarkable acceleration in service has been mainly confined to government and community service, these being the most important service groups which are included in the *A-B* estimates but not in the *C* estimates.

3. *Residual Fluctuation*

The tables of interest are Tables 7 to 13. Table 7 presents the values of $S(x)$, the standard deviation of the residuals, and Table 8 presents the values of $s(x)$, the ratio of $S(x)$ to the average 1947-60 group size. Tables 9 and 10 display the coefficients of correlation of the residuals with the unemployment and total employment residuals, respectively, and Table 11 displays the coefficients of correlation with the "industrial composite" employment residuals (the term "industrial composite" is used here to refer to the total of employment covered in the *C* columns of the tables; it excludes agriculture, fishing and trapping, government, community service, and domestic service). Table 12 expresses the standard deviation of the unemployment residuals as a weighted sum of the standard deviations of the labour force and employment residuals, the weights being the appropriate correlation coefficients as described in Section I, and Table 13 expresse the standard deviation of the total employment residuals as a weighted sum of the standard deviations of the employment residuals in the various industries. It may be noted that the industrial components of the unemployment residuals in Table 12 are obtained by multiplying the standard deviations by the appropriate correlation coefficients and *changing the signs,* this being necessary because the employment residuals appear as negative components in the expression

$$u = f - e_1 - e_2 - \ldots - e_n.$$

In the case of the labour force in Table 12, and in all cases in Table 13, the components retain the signs of their respective correlation coefficients.

The degree of short-run stability or instability in the various series is evident from Tables 7 and 8. (Stability and instability refer, of course, to both cyclical and other non-seasonal types of short-run behaviour.) Unemployment is seen to be a highly unstable series, and employment is seen to be less stable in the aggregate than the total labour force. (Neither of these observations is likely to occasion much surprise!)

Agriculture exhibits a standard deviation which in absolute value is of considerable magnitude but which is relatively small in relation to total employment in this large industry. As one would expect, paid workers account for a disproportionately large share of the instability in agriculture, this being evident from the difference between the figures in the A and B columns of Table 8. Forestry is highly unstable in relation to its total of employment on the evidence of all of the three series examined. The fishing and trapping series exhibits a high degree of relative instability, too, but reference has already been made to the probability that the A and B employment estimates for this small group are subject to considerable sampling variability.

In mining, there is a sharp contrast between the A and B calculations on the one hand, and the C calculations on the other, the former indicating much more instability than the latter. In view of the very high coverage in mining of the indexes on which the C estimates of employment are based, and the possibility of considerable sampling variability in the A and B estimates, it seems almost certain that the C calculations are more reliable in this case.

In manufacturing, taken as a whole, the agreement is extremely good among all of the four series examined, and the standard deviations are seen to be high in terms of absolute numbers but comparatively low in relation to total employment. In general, the durable goods manufacturing industries are much more volatile than the non-durable goods industries, although within the two groups there are several exceptions. The agreement between the C and D calculations, industry by industry, is generally quite good.

In construction, the agreement between the A and B calculations on the one hand, and the C calculations on the other, is not as good as in manufacturing, but the difference is still within reasonable limits.

In most of the service-producing industries there is rather close agreement, the only exception of much consequence being trade, in which the B calculations indicate a much higher degree of instability than the C calculations. In the author's opinion, the C calculations are likely to be more reliable in the case of trade, and if this opinion is accepted the service-producing industries as a group are seen to be relatively much more stable than the goods-producing industries. There are some exceptions, though, most notably communication and public utilities.

In view of the fact that both sampling variability and "real" variability contribute to the movements of the A and B series, it is perhaps not surprising to find that in six of the eight cases in Table 7 where meaningful comparison is possible, and in seven of the eight cases in Table 8, the B calculations give higher results than the C calculations. However, as has been noted, the results differ markedly only in the cases of mining and trade.

Turning next to the correlation coefficients in Tables 9 to 11, it will be noted first that the labour force residuals and the unemployment residuals are practically uncorrelated. On the other hand, the labour force and total employment residuals have a correlation coefficient of 0.64, and this suggests immediately that there has been substantial direct movement between the employed population and the "non-labour-force" population. This provides an interesting starting point for further exploration, although such exploration cannot be undertaken here.

An examination of the detail of the tables reveals a number of cases of differences between the *A* and *B* calculations and the *C* calculations for particular industries. In Table 9, for example, there are notable differences in manufacturing and in construction, the *A* and *B* calculations indicating much less correlation with the unemployment residuals than the *C* calculations. On the other hand, the *C* and *D* calculations are seen once again to agree very well for total manufacturing and reasonably well for most of the components of manufacturing.

Taking the three tables together, and allowing for sampling variability in the *A* and *B* estimates, the evidence suggests that manufacturing, construction, forestry, and transportation have conformed most closely with the general fluctuations of employment and unemployment, and that there has been much less conformity in the service-producing industries other than transportation. It is interesting to note that both in agriculture and in fishing and trapping the employment residuals are *positively* correlated with the unemployment residuals and *negatively* correlated with the total employment residuals. One is tempted to speculate on the possibility that this reflects the tendency for people to remain in, or return to, these industries when the demand for labour elsewhere declines, but the correlation coefficients are not very high and in general the statistical evidence is probably too thin to support much speculation of this kind. There are also some other points of a similar character on which one might wish to speculate (for example, the negative *A* and *B* correlations of the service residuals with the total employment residuals), but again the temptation is resisted and we pass on to the last two tables.

An examination of Table 12 indicates first that fluctuation of the labour force has accounted for a very small proportion of the fluctuation of unemployment— only about 6 per cent. Fluctuation of total employment has accounted for the other 94 per cent.

Because of the substantial differences between the A-B calculations and the *C* calculations, the other conclusions that one can draw from Table 12 are something less than satisfactory, unless one is willing to pass arbitrary judgment on the merits of the different sets of employment data. It is clear that manufacturing, forestry, construction, and transportation have contributed greatly to unemployment fluctuations, and it is clear also that durable goods industries have been responsible for most of manufacturing's contribution, and that the railways have been responsible for a very large part of transportation's contribution. It is difficult to go much beyond this, though, for the calculations disagree very sharply on just how much manufacturing and construction have contributed, and they also disagree very sharply on the contribution of trade. The author's own opinion is that in general the *C* calculations are more reliable measures of short-run fluctuation, but it would not be easy to provide adequate verification of this opinion.

The calculations of Table 13 are generally in very much better agreement than those of Table 12, although it is not clear just why this should be so. In particular, the *A-B* calculations and the *C* calculations indicate roughly the same contributions for manufacturing, construction, and trade. The evidence in Table 13 indicates clearly that forestry, manufacturing (particularly durable goods), construction, transportation (particularly the railways), and trade (particularly retail trade) have accounted for the bulk of the residual fluctuation of total employment. Again the fluctuations in agriculture and in fishing and trapping have tended to offset fluctuations in the other industries, if one can take the calculations at their face value. The *A-B* calculations suggest that service has also behaved in this manner. However, the *C* calculations suggest that in some service industries the fluctuations have contributed to fluctuations of total employment, rather than offsetting them, so that if one is to believe both sets of calculations it must be concluded that the offset has occurred in government and community service, these being the most important groups which are excluded from the *C* estimates of employment. However, the evidence is probably not sufficient to warrant such a conclusion without further investigation.

IV. FINAL NOTE

It will be all too evident to the reader that considerably more work remains to be done before the significance of the results presented in this paper is adequately understood. However, it is hoped that the calculations presented in the tables will provide some useful material for persons concerned with trends and variations in the labour market. If nothing else, they indicate very clearly the dangers of drawing inferences from only one set of employment data. Also, it is hoped that the analytical techniques used in this paper will be of interest both to persons concerned with the analysis of labour statistics and to persons concerned with analysis in other areas of statistics where it is desirable to express movements of totals in terms of the movements of their components.

TABLE 1

Average Group Size (X_m), 1947-1960
(thousands)

	A	B	C	D
Unemployment	232.2	232.2	232.2	232.2
Labour force	5,575.2	5,575.2	5,575.2	5,575.2
Total employment	5,343.0	5,343.0	5,343.0	5,343.0
Employment covered	5,343.0	4,032.9	3,211.9	1,265.5
Agriculture	879.6	112.4	—	—
Forestry	97.4	77.5	99.4	—
Fishing and trapping	26.4	6.7	—	—
Mining	93.4	91.7	104.3	—
Gold mining	—	—	18.5	—
Other metal mining	—	—	39.5	—
Coal mining	—	—	20.2	—
Oil and natural gas	—	—	12.3	—
Other non-metal	—	—	13.8	—
Manufacturing	1,379.4	1,299.1	1,260.9	1,265.5
Durable goods	—	—	630.6	587.5
Wood products	—	—	121.9	127.1
Iron and steel products	—	—	178.7	180.9
Transportation equipment	—	—	174.1	124.3
Non-ferrous metal products	—	—	51.2	50.4
Electrical apparatus and supplies	—	—	70.1	69.7
Non-metallic mineral products	—	—	34.6	35.1
Non-durable goods	—	—	630.3	678.0
Food and beverages	—	—	152.4	180.2
Tobacco and tobacco products	—	—	9.3	10.0
Rubber products	—	—	20.7	21.8
Leather products	—	—	31.7	32.1
Textile products (except clothing)	—	—	65.1	71.4
Clothing (textile and fur)	—	—	97.9	113.4
Paper products	—	—	93.1	85.9
Printing, publishing, etc.	—	—	62.6	67.4
Products of petroleum and coal	—	—	14.8	16.3
Chemical products	—	—	53.3	48.6
Miscellaneous industries	—	—	29.4	30.9
Construction	362.5	296.6	285.9	—
Transportation, storage, and communication	410.2	375.2	379.6	—
Transportation	—	—	301.4*	—
Air transport and airports	—	—	14.2	—
Railways	—	—	169.7	—
Water transportation	—	—	33.2	—
Urban and interurban transportation	—	—	27.6	—
Truck transportation	—	—	42.5	—
Storage	—	—	15.0	—
Communication			63.2	
Public Utilities	59.0	58.8	69.8	—
Trade	802.1	622.3	611.5	—
Wholesale	—	—	191.2	—
Retail	—	—	420.3	—
Finance, insurance, and real estate	174.4	159.7	144.0	—
Service	1.058.6	932.9	356.5†	—
Employment not covered	—	1,310.1	2,131.1	4,077.5

* Includes components for which no separate estimates are available.

† Does not include government, community, or domestic service.

FRANK T. DENTON

TABLE 2

Average Annual Trend Growth (*G*), 1947-1960

	A	B	C	D
Unemployment	23,850	23,850	23,850	23,850
Labour force	106,530	106,530	106,530	106,530
Total employment	82,680	82,680	82,680	82,680
Employment covered	82,680	119,510	62,140	13,460
Agriculture	—36,630	—2,060	—	—
Forestry	270	1,290	—3,970	—
Fishing and trapping	—1,950	—90	—	—
Mining	2,480	2,430	2,460	—
Gold mining	—	—	—530	—
Other metal mining	—	—	2,670	—
Coal mining	—	—	—1,110	—
Oil and natural gas	—	—	1,210	—
Other non-metal	—	—	220	—
Manufacturing	17,190	19,760	13,020	13,460
Durable goods	—	—	8,340	7,820
Wood products	—	—	240	—10
Iron and steel products	—	—	930	2,090
Transportation equipment	—	—	2,380	1,480
Non-ferrous metal products	—	—	1,620	830
Electrical apparatus and supplies	—	—	2,070	2,000
Non-metallic mineral products	—	—	1,100	1,430
Non-durable goods	—	—	4,680	5,640
Food and beverages	—	—	2,210	2,030
Tobacco and tobacco products	—	—	—40	—80
Rubber products	—	—	—60	—120
Leather products	—	—	—510	—410
Textile products (except clothing)	—	—	—1,800	—1,270
Clothing (textile and fur)	—	—	—840	—600
Paper products	—	—	1,790	1,810
Printing, publishing, etc.	—	—	1,280	1,360
Products of petroleum and coal	—	—	470	330
Chemical products	—	—	1,430	1,370
Miscellaneous industries	—	—	750	1,220
Construction	12,810	13,050	7,890	—
Transportation, storage, and communication	5,500	6,250	6,520	—
Transportation	—	—	2,960*	—
Air transport and airports	—	—	990	—
Railways	—	—	—480	—
Water transportation	—	—	—90	—
Urban and interurban transportation	—	—	—430	—
Truck transportation	—	—	2,890	—
Storage	—	—	240	—
Communication	—	—	3,320	—
Public Utilities	3,100	3,150	2,850	—
Trade	27,460	25,230	19,080	—
Wholesale	—	—	5,870	—
Retail	—	—	13,210	—
Finance, insurance, and real estate	7,340	7,120	5,770	—
Service	45,110	43,380	8,520†	—
Employment not covered	—	—36,830	20,540	69,220

* Includes components for which no separate estimates are available.

† Does not include government, community, or domestic service.

TABLE 3

Ratio of Average Annual Trend Growth to Average Group Size (*g*), 1947-1960

	A %	B %	C %	D %
Unemployment	10.27	10.27	10.27	10.27
Labour force	1.91	1.91	1.91	1.91
Total employment	1.55	1.55	1.55	1.55
Employment covered	1.55	2.96	1.93	1.06
Agriculture	—4.16	—1.83	—	—
Forestry	.28	1.66	—3.99	—
Fishing and trapping	—7.39	—1.34	—	—
Mining	2.66	2.65	2.36	—
Gold mining	—	—	—2.86	—
Other metal mining	—	—	6.76	—
Coal mining	—	—	—5.50	—
Oil and natural gas	—	—	9.84	—
Other non-metal	—	—	1.59	—
Manufacturing	1.25	1.52	1.03	1.06
Durable goods	—	—	1.32	1.33
Wood products	—	—	.20	— .01
Iron and steel products	—	—	.52	1.16
Transportation equipment	—	—	1.37	1.19
Non-ferrous metal products	—	—	3.16	1.65
Electrical apparatus and supplies	—	—	2.95	2.87
Non-metallic mineral products	—	—	3.18	4.07
Non-durable goods	—	—	.74	.83
Food and beverages	—	—	1.45	1.13
Tobacco and tobacco products	—	—	— .43	— .80
Rubber products	—	—	— .29	— .55
Leather products	—	—	—1.61	—1.28
Textile products (except clothing)	—	—	—2.76	—1.78
Clothing (textile and fur)	—	—	— .86	— .53
Paper products	—	—	1.92	2.11
Printing, publishing, etc.	—	—	2.04	2.02
Products of petroleum and coal	—	—	3.18	2.02
Chemical products	—	—	2.68	2.82
Miscellaneous industries	—	—	2.55	3.95
Construction	3.53	4.40	2.76	—
Transportation, storage, and communication	1.34	1.67	1.72	—
Transportation	—	—	.98*	—
Air transport and airports	—	—	6.97	—
Railways	—	—	— .28	—
Water transportation	—	—	— .27	—
Urban and interurban transportation	—	—	—1.56	—
Truck transportation	—	—	6.80	—
Storage	—	—	1.60	—
Communication	—	—	5.25	—
Public Utilities	5.25	5.36	4.08	—
Trade	3.42	4.05	3.12	—
Wholesale	—	—	3.07	—
Retail	—	—	3.14	—
Finance, insurance, and real estate	4.21	4.46	4.01	—
Service	4.26	4.65	3.32†	—
Employment not covered	—	—2.81	.96	1.70

* Includes components for which no separate estimates are available.

† Does not include government, community, or domestic service.

TABLE 4

Annual Trend Acceleration (A), 1947-1960

	A	B	C	D
Unemployment	4,670	4,670	4,670	4,670
Labour force	13,150	13,150	13,150	13,150
Total employment	8,480	8,480	8,480	8,480
Employment covered	8,480	4,810	—6,730	—4,240
Agriculture	2,240	530	—	—
Forestry	—130	70	150	—
Fishing and trapping	50	—70	—	—
Mining	—900	—920	—350	—
Gold mining	—	—	10	—
Other metal mining	—	—	—60	—
Coal mining	—	—	—170	—
Oil and natural gas	—	—	—70	—
Other non-metal	—	—	—60	—
Manufacturing	730	150	—4,640	—4,240
Durable goods	—	—	—4,170	—3,690
Wood products	—	—	—250	—580
Iron and steel products	—	—	—460	—460
Transportation equipment	—	—	—2,590	—1,830
Non-ferrous metal products	—	—	—300	—170
Electrical apparatus and supplies	—	—	—580	—640
Non-metallic mineral products	—	—	10	—10
Non-durable goods	—	—	—470	—550
Food and beverages	—	—	50	280
Tobacco and tobacco products	—	—	50	40
Rubber products	—	—	10	—20
Leather products	—	—	60	60
Textile products (except clothing)	—	—	—160	—150
Clothing (textile and fur)	—	—	—300	—260
Paper products	—	—	—70	—100
Printing, publishing, etc.	—	—	—20	—90
Products of petroleum and coal	—	—	—40	—110
Chemical products	—	—	—130	—170
Miscellaneous industries	—	—	80	—30
Construction	—450	—730	—1,380	—
Transportation, storage, and communication	—10	—120	—1,280	—
Transportation	—	—	—1,150*	—
Air transport and airports	—	—	—	—
Railways	—	—	—1,130	—
Water transportation	—	—	—50	—
Urban and interurban transportation	—	—	—60	—
Truck transportation	—	—	130	—
Storage	—	—	—20	—
Communication	—	—	—110	—
Public Utilities	—160	—180	130	—
Trade	—60	—280	—190	—
Wholesale	—	—	—290	—
Retail	—	—	100	—
Finance, insurance, and real estate	520	360	—110	—
Service	6,650	6,000	940†	—
Employment not covered	—	3,670	15,210	12,720

* Includes components for which no separate estimates are available.

† Does not include government, community, or domestic service.

TABLE 5

Ratio of Annual Trend Acceleration to Average Group Size (a), 1947-1960

	A %	B %	C %	D %
Unemployment	2.01	2.01	2.01	2.01
Labour force	.24	.24	.24	.24
Total employment	.16	.16	.16	.16
Employment covered	.16	.12	—.21	—.34
Agriculture	.25	.47	—	— -
Forestry	—.13	.09	.15	—
Fishing and trapping	.19	—1.04	—	—
Mining	—.96	—1.00	—.34	—
Gold mining	—	—	.05	—
Other metal mining	—	—	—.15	—
Coal mining	—	—	—.84	—
Oil and natural gas	—	—	—.57	—
Other non-metal	—	—	—.43	—·
Manufacturing	.05	.01	—.37	—.34
Durable goods	—	—	—.66	—.63
Wood products	—	—	—.21	—.46
Iron and steel products	—	—	—.26	—.25
Transportation equipment	—	—	—1.49	—1.47
Non-ferrous metal products	—	—	—.59	—.34
Electrical apparatus and supplies	—	—	—.83	—.92
Non-metallic mineral products	—	—	.03	—.03
Non-durable goods	—	—	—.07	—.08
Food and beverages	—	—	.03	.16
Tobacco and tobacco products	—	—	.54	.40
Rubber products	—	—	.05	—.09
Leather products	—	—	.19	.19
Textile products (except clothing)	—	—	—.25	—.21
Clothing (textile and fur)	—	—	—.31	—.23
Paper products	—	—	—.08	—.12
Printing, publishing, etc.	—	—	—.03	—.13
Products of petroleum and coal	—	—	—.27	—.67
Chemical products	—	—	—.24	—.35
Miscellaneous industries	—	—	.27	—.10
Construction	—.12	—.25	—.48	—
Transportation, storage, and communication	.00	.03	—.34	—
Transportation	—	—	—.38*	—
Air transport and airports	—	—	.00	—
Railways	—	—	—.67	—
Water transportation	—	—	—.15	—
Urban and interurban transportation	—	—	—.22	—
Truck transportation	—	—	.31	—
Storage	—	—	—.13	—
Communication	—	—	—.17	—
Public Utilities	—.27	—.31	.19	—
Trade	—.01	—.04	—.03	—
Wholesale	—	—	—.15	—
Retail	—	—	.02	—
Finance, insurance, and real estate	.30	.23	—.08	—
Service	.63	.64	.37†	—
Employment not covered	—	.28	.71	.31

* Includes components for which no separate estimates are available.

† Does not include government, community, or domestic service.

FRANK T. DENTON

TABLE 6

Increase or Decrease in Percentage Trend Growth Rate from 1947-48 to 1959-60
(difference in percentage points)

	A	B	C	D
Unemployment	15.32	15.32	15.32	15.32
Labour force	2.38	2.38	2.38	2.38
Total employment	1.62	1.62	1.62	1.62
Employment covered	1.62	.40	—3.08	—4.31
Agriculture	1.07	5.05	—	—
Forestry	—1.67	.85	—.09	—
Fishing and trapping	—4.83	—15.32	—	—
Mining	—13.73	—14.82	—5.12	—
Gold mining	—	—	—.50	—
Other metal mining	—	—	—9.09	—
Coal mining	—	—	—17.17	—
Oil and natural gas	—	—	—33.50	—
Other non-metal	—	—	—5.77	—
Manufacturing	.44	—.14	—4.74	—4.31
Durable goods	—	—	—8.78	—8.38
Wood products	—	—	—.45	—5.77
Iron and steel products	—	—	—3.20	—3.36
Transportation equipment	—	—	—21.49	—21.14
Non-ferrous metal products	—	—	—9.00	—4.71
Electrical apparatus and supplies	—	—	—12.44	—13.88
Non-metallic mineral products	—	—	—.89	—2.46
Non-durable goods	—	—	—.98	—1.02
Food and beverages	—	—	.07	1.67
Tobacco and tobacco products	—	—	5.07	4.78
Rubber products	—	—	.47	—1.45
Leather products	—	—	2.20	2.27
Textile products (except clothing)	—	—	—4.23	—3.02
Clothing (textile and fur)	—	—	—3.89	—2.75
Paper products	—	—	—1.34	—1.99
Printing, publishing, etc.	—	—	—.85	—2.05
Products of petroleum and coal	—	—	—4.90	—9.33
Chemical products	—	—	—4.08	—5.47
Miscellaneous industries	—	—	2.29	—3.11
Construction	—3.19	—5.92	—7.34	—
Transportation, storage, and communication	—.24	—.72	—4.60	—
Transportation	—	—	—4.87*	—
Air transport and airports	—	—	—7.07	—
Railways	—	—	—8.54	—
Water transportation	—	—	—1.85	—
Urban and interurban transportation	—	—	—2.52	—
Truck transportation	—	—	—2.05	—
Storage	—	—	—1.06	--
Communication	—	—	—6.33	--
Public Utilities	—7.42	—8.26	.31	—
Trade	—1.57	—2.70	—1.61	—
Wholesale	—	—	—3.13	—
Retail	—	—	—.92	—
Finance, insurance, and real estate	1.55	.36	—3.04	—
Service	5.47	5.34	3.13†	—
Employment not covered	—	2.45	7.92	3.35

* Includes components for which no separate estimates are available.

† Does not include government, community, or domestic service.

TABLE 7

Standard Deviation of Residuals ($S(x)$), 1947-1960

	A	B	C	D
Unemployment	37,150	37,150	37,150	37,150
Labour force	33,270	33,270	33,270	33,270
Total employment	48,290	48,290	48,290	48,290
Employment covered	48,290	51,670	67,590	25,850
Agriculture	18,590	9,580	—	—
Forestry	12,680	10,610	12,430	—
Fishing and trapping	2,760	1,350	—	—
Mining	8,750	8,820	3,460	—
Gold mining	—	—	770	—
Other metal mining	—	—	1,840	—
Coal mining	—	—	1,020	—
Oil and natural gas	—	—	900	—
Other non-metal	—	—	610	—
Manufacturing	25,770	24,070	28,060	25,850
Durable goods	—	—	23,480	21,360
Wood products	—	—	3,000	2,890
Iron and steel products	—	—	7,680	7,320
Transportation equipment	—	—	14,170	10,050
Non-ferrous metal products	—	—	1,610	1,620
Electrical apparatus and supplies	—	—	3,290	2,831
Non-metallic mineral products	—	—	900	940
Non-durable goods	—	—	7,670	6,750
Food and beverages	—	—	1,440	2,040
Tobacco and tobacco products	—	—	330	310
Rubber products	—	—	970	930
Leather products	—	—	780	690
Textile products (except clothing)	—	—	3,510	3,170
Clothing (textile and fur)	—	—	3,170	2,230
Paper products	—	—	2,180	1,380
Printing, publishing, etc.	—	—	820	940
Products of petroleum and coal	—	—	340	200
Chemical products	—	—	1,000	1,040
Miscellaneous industries	—	—	980	910
Construction	18,720	16,650	13,210	—
Transportation, storage, and communication	11,400	10,840	9,390	—
Transportation	—	—	7,370*	—
Air transport and airports	—	—	740	—
Railways	—	—	5,230	—
Water transportation	—	—	1,180	—
Urban and interurban transportation	—	—	620	—
Truck transportation	—	—	1,170	—
Storage	—	—	520	—
Communication	—	—	2,890	—
Public Utilities	2,360	2,360	2,300	—
Trade	22,550	19,900	8,630	—
Wholesale	—	—	2,650	—
Retail	—	—	6,500	—
Finance, insurance, and real estate	3,600	2,960	1,930	—
Service	12,250	12,540	4,320†	—
Employment not covered	—	14,880	30,680	31,390

* Includes components for which no separate estimates are available.

† Does not include government, community, or domestic service.

TABLE 8

Ratio of Standard Deviation of Residuals to Average Group Size ($s(x)$), 1947-1960

	A %	B %	C %	D %
Unemployment	16.0	16.0	16.0	16.0
Labour force	.60	.60	.60	.60
Total employment	.90	.90	.90	.90
Employment covered	.90	1.28	2.10	2.04
Agriculture	2.11	8.52	—	—
Forestry	13.02	13.69	12.51	—
Fishing and trapping	10.45	20.15	—	—
Mining	9.37	9.62	3.32	—
Gold mining	—	—	4.16	—
Other metal mining	—	—	4.66	—
Coal mining	—	—	5.05	—
Oil and natural gas	—	—	7.32	—
Other non-metal	—	—	4.42	—
Manufacturing	1.87	1.85	2.23	2.04
Durable goods	—	—	3.72	3.64
Wood products	—	—	2.46	2.27
Iron and steel products	—	—	4.30	4.05
Transportation equipment	—	—	8.14	8.09
Non-ferrous metal products	—	—	3.14	3.21
Electrical apparatus and supplies	—	—	4.69	4.06
Non-metallic mineral products	—	—	2.60	2.68
Non-durable goods	—	—	1.22	1.00
Food and beverages	—	—	.94	1.13
Tobacco and tobacco products	—	—	3.55	3.10
Rubber products	—	—	4.69	4.27
Leather products	—	—	2.46	2.15
Textile products (except clothing)	—	—	5.39	4.44
Clothing (textile and fur)	—	—	3.24	1.97
Paper products	—	—	2.34	1.61
Printing, publishing, etc.	—	—	1.31	1.39
Products of petroleum and coal	—	—	2.30	1.23
Chemical products	—	—	1.88	2.14
Miscellaneous industries	—	—	3.33	2.94
Construction	5.16	5.61	4.62	—
Transportation, storage, and communication	2.78	2.89	2.47	—
Transportation	—	—	2.45*	—
Air transport and airports	—	—	5.21	—
Railways	—	—	3.08	—
Water transportation	—	—	3.55	—
Urban and interurban transportation	—	—	2.25	—
Truck transportation	—	—	2.75	—
Storage	—	—	3.47	—
Communication	—	—	4.57	—
Public Utilities	4.00	4.01	3.30	—
Trade	2.81	3.20	1.41	—
Wholesale	—	—	1.39	—
Retail	—	—	1.55	—
Finance, insurance, and real estate	2.06	1.85	1.34	—
Service	1.16	1.34	1.68†	—
Employment not covered	—	1.14	1.44	.77

* Includes components for which no separate estimates are available.

† Does not include government, community, or domestic service.

TABLE 9

Coefficients of Correlation of Residuals with Unemployment Residuals, 1947-1960

	A	B	C	D
Unemployment	1.00	1.00	1.00	1.00
Labour force	.06	.06	.06	.06
Total employment	—.73	—.73	—.73	—.73
Employment covered	—.73	—.77	—.74	—.85
Agriculture	.41	.17	—	—
Forestry	—.64	—.59	—.67	—
Fishing and trapping	.28	.25	—	—
Mining	—.14	—.15	—.24	—
Gold mining	—	—	.27	—
Other metal mining	—	—	—.32	—
Coal mining	—	—	.19	—
Oil and natural gas	—	—	—.33	—
Other non-metal	—	—	—.58	—
Manufacturing	—.46	—.52	—.85	—.85
Durable goods	—	—	—.82	—.82
Wood products	—	—	—.56	—.77
Iron and steel products	—	—	—.80	—.88
Transportation equipment	—	—	—.52	—.48
Non-ferrous metal products	—	—	—.84	—.86
Electrical apparatus and supplies	—	—	—.71	—.72
Non-metallic mineral products	—	—	—.64	—.48
Non-durable goods	—	—	—.58	—.66
Food and beverages	—	—	—.44	—.22
Tobacco and tobacco products	—	—	.46	.32
Rubber products	—	—	—.73	—.80
Leather products	—	—	—.42	—.36
Textile products (except clothing)	—	—	—.30	—.31
Clothing (textile and fur)	—	—	—.15	—.25
Paper products	—	—	—.60	—.66
Printing, publishing, etc.	—	—	.06	—.38
Products of petroleum and coal	—	—	—.22	—.16
Chemical products	—	—	—.35	—.22
Miscellaneous industries	—	—	.28	—.04
Construction	—.22	—.23	—.57	—
Transportation, storage, and communication	—.67	—.64	—.63	—
Transportation	—	—	—.76*	—
Air transport and airports	—	—	—.37	—
Railways	—	—	—.75	—
Water transportation	—	—	—.48	—
Urban and interurban transportation	—	—	.33	—
Truck transportation	—	—	—.64	—
Storage	—	—	—.40	—
Communication	—	—	—.06	—
Public Utilities	.18	.17	.23	—
Trade	—.46	—.50	—.35	—
Wholesale	—	—	—.58	—
Retail	—	—	—.22	—
Finance, insurance, and real estate	—.36	—.26	—.05	—
Service	.05	—.05	—.20†	—
Employment not covered	—	.31	.48	—.42

* Includes components for which no separate estimates are available.

† Does not include government, community, or domestic service.

TABLE 10

Coefficients of Correlation of Residuals with Total Employment Residuals, 1947-1960

	A	B	C	D
Unemployment	—.73	—.73	—.73	—.73
Labour force	.64	.64	.64	.64
Total employment	1.00	1.00	1.00	1.00
Employment covered	1.00	.96	.91	.81
Agriculture	—.30	—.18	—	—
Forestry	.40	.49	.42	—
Fishing and trapping	—.08	—.03	—	—
Mining	.54	.54	.57	—
Gold mining	—	—	—.20	—
Other metal mining	—	—	.67	—
Coal mining	—	—	—.06	—
Oil and natural gas	—	—	.79	—
Other non-metal	—	—	.38	—
Manufacturing	.78	.81	.85	.81
Durable goods	—	—	.79	.71
Wood products	—	—	.33	.29
Iron and steel products	—	—	.76	.80
Transportation equipment	—	—	.55	.53
Non-ferrous metal products	—	—	.65	.63
Electrical apparatus and supplies	—	—	.71	.62
Non-metallic mineral products	—	—	.52	.30
Non-durable goods	—	—	.70	.86
Food and beverages	—	—	.56	.69
Tobacco and tobacco products	—	—	.04	.18
Rubber products	—	—	.53	.51
Leather products	—	—	.49	.52
Textile products (except clothing)	—	—	.26	.30
Clothing (textile and fur)	—	—	.20	.33
Paper products	—	—	.57	.69
Printing, publishing, etc.	—	—	.22	.65
Products of petroleum and coal	—	—	.66	.16
Chemical products	—	—	.65	.25
Miscellaneous industries	—	—	—.24	—.05
Construction	.63	.67	.82	—
Transportation, storage, and communication	.62	.61	.93	—
Transportation	—	—	.90*	—
Air transport and airports	—	—	.44	—
Railways	—	—	.87	—
Water transportation	—	—	.71	—
Urban and interurban transportation	—	—	—.22	—
Truck transportation	—	—	.66	—
Storage	—	—	.68	—
Communication	—	—	.61	—
Public Utilities	.36	.39	.36	—
Trade	.30	.31	.79	—
Wholesale	—	—	.94	—
Retail	—	—	.66	—
Finance, insurance, and real estate	.85	.78	.13	—
Service	—.44	—.43	.76†	—
Employment not covered	—	—.08	—.44	.87

* Includes components for which no separate estimates are available.

† Does not include government, community, or domestic service.

TABLE 11

Coefficients of Correlation of Residuals with "Industrial Composite" Employment Residuals, 1947-1960

	A	B	C	D
Unemployment	—.74	—.74	—.74	—.74
Labour force	.50	.50	.50	.50
Total employment	.91	.91	.91	.91
Employment covered	.91	.96	1.00	.86
Agriculture	—.61	—.46	—	—
Forestry	.51	.55	.62	—
Fishing and trapping	—.25	.04	—	—
Mining	.30	.30	.72	—
Gold mining	—	—	—.09	—
Other metal mining	—	—	.79	—
Coal mining	—	—	.00	—
Oil and natural gas	—	—	.82	—
Other non-metal	—	—	.60	—
Manufacturing	.75	.79	.93	.86
Durable goods	—	—	.85	.78
Wood products	—	—	.44	.39
Iron and steel products	—	—	.88	.91
Transportation equipment	—	—	.58	.59
Non-ferrous metal products	—	—	.57	.68
Electrical apparatus and supplies	—	—	.67	.61
Non-metallic mineral products	—	—	.60	.21
Non-durable goods	—	—	.79	.82
Food and beverages	—	—	.70	.57
Tobacco and tobacco products	—	—	—.14	.03
Rubber products	—	—	.54	.60
Leather products	—	—	.41	.34
Textile products (except clothing)	—	—	.38	.39
Clothing (textile and fur)	—	—	.30	.34
Paper products	—	—	.53	.65
Printing, publishing, etc.	—	—	.17	.44
Products of petroleum and coal	—	—	.80	.06
Chemical products	—	—	.64	.22
Miscellaneous industries	—	—	—.24	.01
Construction	.59	.62	.90	—
Transportation, storage, and communication	.79	.80	.94	—
Transportation	—	—	.94*	—
Air transport and airports	—	—	.25	—
Railways	—	—	.91	—
Water transportation	—	—	.70	—
Urban and interurban transportation	—	—	—.24	—
Truck transportation	—	—	.82	—
Storage	—	—	.68	—
Communication	—	—	.55	—
Public Utilities	.40	.43	.12	—
Trade	.32	.36	.76	—
Wholesale	—	—	.95	—
Retail	—	—	.62	—
Finance, insurance, and real estate	.80	.75	.41	—
Service	—.26	—.24	.71†	—
Employment not covered	—	—.37	—.77	.70

* Includes components for which no separate estimates are available.

† Does not include government, community, or domestic service.

FRANK T. DENTON

TABLE 12

Contributions to Residual Fluctuation of Unemployment (S(x) Weighted by Coefficient of Correlation with Unemployment Residuals), 1947-1960*

	A	B	C	D
Unemployment	37,150	37,150	37,150	37,150
Labour force	2,090	2,090	2,090	2,090
Total employment	35,060	35,060	35,060	35,060
Employment covered	35,060	39,630	49,870	21,870
Agriculture	—7,700	—1,620	—	—
Forestry	8,050	6,300	8,280	—
Fishing and trapping	—770	—330	—	—
Mining	1,270	1,300	850	—
Gold mining	—	—	—210	—
Other metal mining	—	—	600	—
Coal mining	—	—	—190	—
Oil and natural gas	—	—	290	—
Other non-metal	—	—	360	—
Manufacturing	11,720	12,400	23,800	21,850
Durable goods	—	—	19,370	17,410
Wood products	—	—	1,690	2,230
Iron and steel products	—	—	6,120	6,420
Transportation equipment	—	—	7,300	4,870
Non-ferrous metal products	—	—	1,350	1,390
Electrical apparatus and supplies	—	—	2,330	2,050
Non-metallic mineral products	—	—	580	450
Non-durable goods	—	—	4,430	4,440
Food and beverages	—	—	640	440
Tobacco and tobacco products	—	—	—150	—100
Rubber products	—	—	710	740
Leather products	—	—	320	250
Textile products (except clothing)	—	—	1,040	990
Clothing (textile and fur)	—	—	470	560
Paper products	—	—	1,300	900
Printing, publishing, etc.	—	—	—50	360
Products of petroleum and coal	—	—	80	30
Chemical products	—	—	350	230
Miscellaneous industries	—	—	—280	40
Construction	4,140	3,850	7,560	—
Transportation, storage, and communication	7,630	6,880	5,960	—
Transportation	—	—	5,590†	—
Air transport and airports	—	—	280	—
Railways	—	—	3,940	—
Water transportation	—	—	560	—
Urban and interurban transportation	—	—	—200	—
Truck transportation	—	—	750	—
Storage	—	—	210	—
Communication	—	—	160	—
Public Utilities	—440	—400	—540	—
Trade	10,460	9,910	2,990	—
Wholesale	—	—	1,550	—
Retail	—	—	1,440	—
Finance, insurance, and real estate	1,280	780	100	—
Service	—580	560	870‡	—
Employment not covered	—	—4,570	—14,810	13,210

* Since employment and unemployment are inversely related, the contributions of employment are obtained by multiplying S(x) by the coefficients of correlation *with signs changed.*

† Includes components for which no separate estimates are available.

‡ Does not include government, community, or domestic service.

TABLE 13

Contributions to Residual Fluctuation of Total Employment ($S(x)$ Weighted by Coefficient of Correlation with Total Employment Residuals), 1947-1960

	A	B	C	D
Total employment	48,290	48,290	48,290	48,290
Employment covered	48,290	49,500	61,710	20,860
Agriculture	—5,650	—1,740	—	—
Forestry	5,090	5,230	5,250	—
Fishing and trapping	—230	—40	—	—
Mining	4,690	4,740	1,960	—
Gold mining	—	—	—150	—
Other metal mining	—	—	1,230	—
Coal mining	—	—	—60	—
Oil and natural gas	—	—	710	—
Other non-metal	—	—	230	—
Manufacturing	20,230	19,500	23,800	20,860
Durable goods	—	—	18,460	15,080
Wood products	—	—	1,000	850
Iron and steel products	—	—	5,800	5,830
Transportation equipment	—	—	7,810	5,340
Non-ferrous metal products	—	—	1,050	1,020
Electrical apparatus and supplies	—	—	2,330	1,760
Non-metallic mineral products	—	—	470	280
Non-durable goods	—	—	5,340	5,780
Food and beverages	—	—	810	1,420
Tobacco and tobacco products	—	—	10	60
Rubber products	—	—	520	470
Leather products	—	—	380	360
Textile products (except clothing)	—	—	910	930
Clothing (textile and fur)	—	—	640	730
Paper products	—	—	1,240	960
Printing, publishing, etc.	—	—	180	610
Products of petroleum and coal	—	—	230	30
Chemical products	—	—	660	260
Miscellaneous industries	—	—	—240	—50
Construction	11,850	11,190	10,810	—
Transportation, storage, and communication	7,060	6,640	8,750	—
Transportation	—	—	6,630*	—
Air transport and airports	—	—	320	—
Railways	—	—	4,530	—
Water transportation	—	—	840	—
Urban and interurban transportation	—	—	—140	—
Truck transportation	—	—	770	—
Storage	—	—	350	—
Communication	—	—	1,770	—
Public Utilities	860	920	820	—
Trade	6,660	6,110	6,810	—
Wholesale	—	—	2,490	—
Retail	—	—	4,320	—
Finance, insurance, and real estate	3,060	2,320	250	—
Service	—5,330	—5,370	3,260†	—
Employment not covered	—	—1,210	—13,420	27,430

Inter-Industry Estimates of Canadian Imports 1949-1958*

T. I. MATUSZEWSKI

PAUL R. PITTS

JOHN A. SAWYER

TABLES OF THE INTER-INDUSTRY FLOW of goods and services have become widely accepted as a useful addition to the system of national economic account. When certain averaging assumptions are made, such tables enable the analyst to make quantitative statements for the year for which the table was constructed; for example, the import content of personal expenditure on consumer goods and services purchased by Canadians in 1949 was 16 per cent, while the import content of business gross fixed capital formation was 27 per cent [6, Table 4].[1] Obviously, it would be useful if an input-output model could be developed which would predict the effects of expected or forecasted changes in final demand.

A prime condition for reliable input-output predictions is that the parameters of the model remain sufficiently stable over time. For variables that can be predicted by other methods, the predictions made by the input-output model should be more reliable than those made by alternative techniques of equal or lesser cost, and the value of the gain in precision of the input-output technique should exceed the additional cost of the technique. If these conditions are satisfied, then a useful tool for economic analysis will have been developed. It should be emphasized, however, that input-output analysis is designed to answer questions not readily answered by other techniques and, when used for this purpose, complements, rather than competes with, these techniques.

To obtain estimates of the parameters of an input-output model, a table of the inter-industry flow of goods and services should be constructed in such a way that it can be adapted to the specifications of the model. Input-output models used by various researchers differ, however, in their assumptions concerning the

* Revisions based on the oral presentation and discussion of the paper in Montreal and Kingston have been made to the original draft circulated to the Conference. Port of the work for this paper was done at the Institute for Economic Research, Queen's University, Kingston, Canada. Financial support was also provided by the Canada Council and the Institute of Social and Economic Research of The University of British Columbia. The inverse matrices were calculated at the Computing Centre of the University of British Columbia. The authors also wish to express their indebtedness to Mr. A. A. Tooms of the Dominion Bureau of Statistics for preparing the current dollar estimates of competitive and non-competitive imports and to other members of D.B.S. who assisted in making this study possible.

[1] Numbers in [ ] refer to references listed at the end of the paper.

C.P.S.A. Conference on Statistics. 1961, *Papers*. Printed in the Netherlands.

relation between imports and domestic production, as wel as in other respects. The authors of this paper began in 1958 a systematic exploration of the various treatments of imports in input-output models and this paper is a progress report on the quantitative aspects of their findings. It is hoped that the findings will throw light on the question of whether, and for how long, input-output relations remain sufficiently stable to make reliable predictions possible. If there is stability in some of the various relations, it is further hoped that the findings will be of guidance to the Dominion Bureau of Statistics in the construction of further tables of the inter-industry flow of good and services, in deciding what information should be shown in the table, and what supplementary information should be published with it, including the appropriate inverse matrices.

In this paper, the changes in Canadian final demand during the period 1949-58 are examined at various levels of detail and the effects of these changes "backcast" through to the implied effects upon the intermediate output of Canadian industries and upon intermediate imports. The backcasts are accomplished using three variations of the 1949 table of the inter-industry flow of goods and services at producers' prices [6, Table 1], each variation incorporating a different relation between imports and domestic output. The backcasts are then compared with estimated "actual figures to measure the errors of input-output projections, and compared with estimates derived by other techniques to assess the efficiency of input-output analysis. One of the major problems facing input-output analysis is the specification of the final demand in appropriate detail and classification for the input-output model. Until this problem is solved, it will be difficult to assess properly the true value of input-output analysis.

I. IMPORTS AND DOMESTIC PRODUCTION

Input-output coefficients, the ratio of the amount of an input used in an industry to the output of the industry, are a composite resulting from both technical and marketing relations. In this paper, the hypothesis is tested that technical relations tend to remain constant while instability results from shifts in sources of supply. Two types of alternative sources of supply exist. The first is where the same commodity is produced by more than one domestic industry—the problem of secondary products. The forty-two industry flow table constructed by D.B.S. [6] is an aggregation of industries in such a way that this is a minor problem. Moreover, some of the analysis has been done with a sixteen-industry aggregation where the problem virtually disappears. The alternative of substituting imported goods for domesticaly-produced goods is, however, present regardless of the level of aggregation

Three different input-output models have been developed which incorporate three different relations between imports and domestic production. The models are not new; each of them has been used at various times by various researchers. In their simplest form, these models assume that each domestic industry produces one had only one homogeneous commodity, so that there is a one-to-one correspondence between industries and commodities. It is als assumed that a change in the level of output of an industry requires that all commodity inputs (including non-factor services) change in the same proportion; that is, average and marginal input coefficients are identical. The models differentiate between imports of a kind not produced in Canada ("non-competitive" or "complementary" imports) and imports of a kind produced in Canada ("competitive" or "supplementary" imports). In all three models, non-competitive imports are treated as primary inputs into the industry which uses them. Competitive imports are treated differently, however, in each of the three models.

In this paper the complete system of commodity balance equations which comprise an input-output system will not be reproduced since they may be found in another paper by the present author [51]. Instead, attention will be focused on the assumptions about the stability of parameters which constitute the essential differences among the three models. The following notation will be used in defining these parameters:

Flows

X_{ij}^D — domestic output of commodity i used as input by industry j (intermediate output of industry i)
$i = 1, 2, \ldots, n$ refers to commodities
$j = 1, 2, \ldots, n$ refers to the activity of producing or importing a commodity
There is a one-to-one correspondence between commodities and activities.

X_{ij}^M — imports of commodity i used as input by industry j (intermediate competitive imports of commodity i)

$X_{ij} = (X_{ij}^D + X_{ij}^M)$ — total quantity of commodity i used by industry j

X_j^D — level of activity of industry j = quantity of domestic output of commodity j

X_j^M — level of importing activity j = quantity of imports of commodity j

$X_i = (X_i^D + X_i^M)$ — total supply of commodity i

Y_i^D — final demand for domestic output of commodity i

Y_i^M — final demand for imports of commodity i = predetermined imports of commodity i

$Y_i = (Y_i^D + Y_i^M)$ — total or "combined" final demand for commodity i

Parameters

$a_{ij} = \dfrac{X_{ij}^D}{X_j^D}$ — domestic input coefficient

$m_{ij} = \dfrac{X_{ij}^M}{X_j^D}$ — import input coefficient

$(a_{ij} + m_{ij}) = \dfrac{(X_{ij}^D + X_{ij}^M)}{X_j^D}$ — total or "combined" input coefficient. When written in this form the coefficients a_{ij} and m_{ij} have no seperate existence; only their sum is assumed to be known.

$\dfrac{m_i}{1 + m_i}$ — import share of the market for commodity i, where $m_i = \dfrac{X_i^M}{X_i^D}$

$\dfrac{1}{1 + m_i}$ — domestic share of the market for commodity i.

In Model I, no substitution is permitted between domestic output and competitive imports. In effect, therefore, all imports are treated as non-competitive. The input coefficients for the domestic commodities used by an industry (a_{ij}) and the coefficients for imported commodities used by the industry (m_{ij}) are distinct and each is assumed to be a constant. The level of output of a domestic industry (X_j^D) determines the level of imports used in that industry (X_{ij}^M). There is no influence in the opposite direction since imports do not "compete" with domestic output. Direct final demand for imported commodities (Y_i^M) is regarded as predetermined. This is the model underlying the inter-industry flow tables for 1949 published by D.B.S. [5 and 6].

In Model II, perfect substitution between imports and domestic output is assumed. Thus, an input into a particular industry may be either domestically produced, imported, or a combination of the two. It is this aggregate input coefficient (the total input per unit of output, regardless of source, $a_{ij}+m_{ij}$) that is assumed to be constant. Substitution of imports for domestic output will not change this coefficient, whereas in Model I such substitution changes two coefficients. Competitive imports are routed into the system of inter-industry flows as an input into the domestic industry which produces the same commodity. It is assumed in this model that the ratio of the imports of a competitive commodity to the output of the competitive domestic industry (m_i) is a constant in constant dollars; that is, the foreign and domestic shares of the market $(m_i/(1+m_i))$ and $(1/(1+m_i))$ remain constant. In this model no competitive imports are imported directly by final demand sectors; all such imports are routed through the competing domestic industry. This model has been used, for example, by Chenery, Clark, and Cao Pinna in their study of the Italian economy [48].

Model III represents a retreat from the objective of constructing a model which explains the level of competitive imports. In this model the same assumptions as in Model II with respect to the substitutability of domestic commodities and competitive imports and constancy of the aggregate input coefficients into an industry $(a_{ij}+m_{ij})$ are made. However, the Model II assumption of fixed shares of domestic industries and imports in markets for competitive commodities (i.e., a constant m_i) is dropped. The level of competitive imports (X_i^M) must therefore be are determined by factors other than those incorporated into the model.[2] Model III therefore comes closest to representing "true" technology. Although the model does not explain the level of competitive imports, it does, as do the other two models, explain the levels of intermediate output of domestic industries $(X_i^D - Y_i^D)$ and of intermediate non-competitive imports. Model III was used by Arrow and Hoffenberg [12, p. 75] and by the United States Bureau of Labor Statistics for the 1947 "emergency model" of the United States [cited in 18, p. 154]. The detailed B.L.S. study for 1947 was based on Model II [24, p. 109].

Flow tables for Canada for 1949 at producers' prices have been constructed ac-

---

[2] Various interpretations of treating competitive imports as predetermined in an input-output model are possible. One is that some other model is used to predict their level and the results are then fed into the input-output model as a given. A second is that there are direct import controls which determine their (maximum) level. A third is that a system of tariffs or subsidies has been designed so that a predetermined level of competitive imports is obtained.

cording to all three models at the forty-two industry level of detail. Flow Table I
was published by D.B.S. in [6] and, on the basis of it and estimates of competitive
imports cross-classified by using and by competing industry, Flow Tables II and III
were constructed (but have not been published).[3] Inverse matrices were then calculated for all three models. Sixteen-industry aggregations of the three forty-two
industry flow tables were also made and their inverse matrices calculated.[4] Results
are given in many cases for both aggregations.

II. FINAL DEMAND, INTERMEDIATE OUTPUT, AND IMPORTS

As a result of the different treatment of competitive imports in each of the three
models, each model defines final demand in a different manner. In Model I, the final
demand for each commodity must be separated into demand for the product of the
domestic industry (Y_i^D) and demand for the imported commodity (Y_i^M). The latter is
directed outside the model and has no effect on domestic activity. The demand
directed towards domestic industries then determines the levels of intermediate output
of those industries through the network of inter-industry relations summarized in the
inverse of the matrix comprised of an identity matrix less the matrix of input coefficients. The competitive imports used in an industry (X_{ij}^m) can then be determined by
applying the ratio of imports of each commodity to the output of the industry in the
base year (m_{ij}) to the projected level of output (X_i^D) for the year for which the final
output forecast has been made. Similarly the quantity of non-competitive imports
and any other primary inputs can be determined.[5]

The presence of unallocated output in the D.B.S. 1949 Table complicates the specification of final demand. Rather than incorporate unallocated inputs and outputs into
the inverse matrix, they were excluded. In order to preserve the fundamental identity
that a dollar's worth of final demand gives rise to a dollar's worth of primary input
somewhere in the system, unallocated output was treated as if it were a final demand
and unallocated input as if it were a primary input. In Model I, therefore, total final
demand for the output of an industry consists of the final demand for the output of
the domestic industry plus the unallocated output of that industry. For this paper
it was assumed that unallocated output in all years was the same amount as in 1949.

In Model II the production of a domestic industry (X_i^D) is supplemented by the

---

[3] The flow table in [5] is at purchasers' prices and was published prior to the revision of the D.B.S.
National Accounts [4].

[4] Copies of the sixteen-industry aggregation corresponding to Model I and its inverse matrix
are available upon request from Mr. P. R. Pitts, Central Research and Development Staff, Dominion
Bureau of Statistics, Ottawa. Copies of an eight-industry table aggregated on different criteria and
its inverse are also available.

[5] A brief mathematical note on the derivation of the primary input content of final output for
Model I is given in [6, p. 18]. It should be noted that the inverse matrices used in this project exclude
commodities produced and used within the same industry. The output figures derived from the
analysis, therefore, exclude intra-industry consumption and must be raised to include this production
if they are to be comparable with industry output figures from other sources. The multipliers for the
forty-two-industry table, Model I, are given in [6, p. 12, n. 13].

addition of imports of that product (X_i^m). It is only necessary in this case to specify total final demand for a commodity, $Y_i = (Y_i^D + Y_i^M)$; it is not necessary to specify separately the domestically and foreign-produced quantities. Model II allocates the two sources of supply in the same proportions of the total market which they had in the base year, 1949 (that is, m_i is assumed stable). Model II therefore backcasts the level of total output of the domestic industry and the level of competitive imports of that industry, without specifying the proportion of each which was used in a particular industry.

In Model III, all imports are predetermined and the final demand directed towards an industry is the total final demand for that commodity (as in Model II) *less* all imports of that commodity whether for final or intermediate use ($Y_i - X_i^m$). (Model I only deducted imports for final use.) Model III, therefore, requires knowledge of total final demand for a commodity, regardless of whether it is directed towards domestic or foreign sources, and knowledge of total imports of the commodity, Model II thus requires the least information and Model I the most.

The three models are summarized in Table 1.

TABLE 1

Summary of the Three Input-Output Models

	Final demand to be specified	Parameters assumed stable	Activity levels to be predicted
Model I	$Y_i^D = (Y_i - Y_i^M)$	a_{ij} m_{ij}	$(X_j^D - Y_j^D)$ $(X_j^M - Y_j^M)$
Model II	Y_i	$(a_{ij} + m_{ij})$ m_i	X_j^D X_j^M
Model III	$(Y_i - X_i^M)$	$(a_{ij} + m_{ij})$	X_j^D

III. BACKCASTS FROM 1956 FINAL DEMAND

For this project a detailed study was made of 1956 final demand for the output of the forty-two industries of [6] by eleven categories of final demand in 1949 dollars. Tables 2 and 3 list the forty-two-industries and the sixteen-industry aggregation. The starting point was the constant dollar expenditure data in the National Accounts [4, Tables 5 and 48] and [3 (1959), Tables 5 and 48]. These were supplemented by unpublished expenditure estimates at a finer level of detail from the worksheets of the National Accounts Division at D.B.S. and by worksheet data on inventories from the Inventories Section, Industry and Merchandising Division and on constant dollar exports and imports from the Exteral Trade Section, International Trade Division of D.B.S. Lack of resources made commodity flow studies of the disposition of output of each industry impossible. When expenditure detail was not sufficient to determine the industry of origin, 1949 ratios were used. The direct import content of final demand was determined from [10] and converted into 1949 dollars using import price indexes.

TABLE 2

The Forty-Two Industries

Industry Number	Industry	Industry Number	Industry
1	Agriculture	22	Furniture
2	Forestry	23	Wood products (except furniture)
3	Fishing, hunting and trapping	24	Paper products
4	Metal mining and smelting and re-fining	25	Printing, publishing, and allied industries
5	Coal mining, crude petroleum and natural gas	26	Primary iron and steel
6	Non-metal mining, quarrying, and prospecting	27	Agricultural implements
		28	Iron and steel products, n.e.s.
7	Meat products	29	Transportation equipment
8	Dairy products	30	Jewellery and silverware (incl. watch repair)
9	Fish processing	31	Non-ferrous metal products, n.e.s.
10	Fruit and vegetable preparations	32	Electrical apparatus and supplies
11	Grain mill products	33	Non-metallic mineral products
12	Bakery products	34	Products of petroleum and coal
13	Carbonated beverages	35	Chemicals and allied products
14	Alcoholic beverages	36	Miscellaneous manufacturing industries
15	Confectionery and sugar refining	37	Construction
16	Miscellaneous food preparations	38	Transportation, storage, and trade
17	Tobacco and tobacco products	39	Communication
18	Rubber products	40	Electric power, gas, and water utilities
19	Leather products	41	Finance, insurance, and real estate
20	Textile products (except clothing)	42	Service industries
21	Clothing (textile and fur)		

Note: For the relations of these industries to the D.B.S. Standard Industrial Classification (1948), see [6, Table 10 and pages 26-34]. "N.E.S." means "not elsewhere specified."

TABLE 3

Aggregation of Industries for Sixteen-Industry Table

Sixteen-Industry Table		Forty-Two-Industry Table	
Industry Number	Name of Industry	Industry Number	Name of Industry
1	Agriculture	1	Agriculture
2	Forestry and fishing	2	Forestry
		3	Fishing, hunting and trapping
3	Mining, smelting and refining	4	Metal mining and smelting and refining
		6	Non-metal mining, quarrying, and prospecting
4	Coal mining, crude petroleum and natural gas	5	Coal mining, crude petroleum and natural gas

TABLE 3 (*continued*)

Sixteen-Industry Table		Forty-Two-Industry Table	
Industry Number	Name of Industry	Industry Number	Name of Industry
5	Foods, beverages, and tobacco	7	Meat products
		8	Dairy products
		9	Fish processing
		10	Fruit and vegetable preparations
		11	Grain mill products
		12	Bakery products
		13	Carbonated beverages
		14	Alcoholic beverages
		15	Confectionery and sugar refining
		16	Miscellaneous food preparations
		17	Tobacco and tobacco products
6	Leather and textile products, clothing	19	Leather products
		20	Textile products (except clothing)
		21	Clothing (textile and fur)
7	Wood products	22	Furniture
		23	Wood products (except furniture)
8	Paper products	24	Paper products
9	Iron and steel and non-ferrous metal products	26	Primary iron and steel
		28	Iron and steel products, n.e.s.
		31	Non-ferrous metal products, n.e.s.
10	Transportation equipment and electrical apparatus	27	Agricultural implements
		29	Transportation equipment
		32	Electrical apparatus and supplies
11	Chemical, rubber, petroleum and coal products	18	Rubber products
		34	Products of petroleum and coal
		35	Chemicals and allied products
12	Other manufacturing industries	25	Printing, publishing and allied industries
		30	Jewellery and silverware (incl., watch repair)
		33	Non-metallic mineral products
		36	Miscellaneous manufacturing industries
13	Construction	37	Construction
14	Transportation, storage, and trade	38	Transportation, storage, and trade
15	Electric power, gas, and water utilities	40	Electric power, gas, and water utilities
16	Other service industries	39	Communication
		41	Finance, insurance, and real estate
		42	Service industries

Unallocated final output was assumed to be the same as in 1949. These estimates of final demand for Model I are shown in Table 4 and are comparable to the figures in columns 43 and 45-50 of Table 1 of [6]. Table 5 shows how the total final demand for each industry is converted into final demand for Models II and III, respectively.

TABLE 4

Final Demand Directed Towards Domestic Industries, by Expenditure Categories 1956, Model I
(millions of 1949 dollars)

Industry number	1 Uuallocated output	2 Personal expenditure on consumer goods and services	3 Government expenditure on goods and services	4 Business gross fixed capital formation	5 Value of physical change in inventories	6 Exports of goods and services	7 Residual error of estimate	8 Total final demand
1		867		8	307	789		1,971
2		6			7	69		82
3		11			7	52		70
4	1				42	834		877
5		30			9	102		141
6	4				6	76		86
7	4	776				100		880
8	19	474			1	35		529
9	1	50			1	75		127
10	2	193			6	8		209
11		96			10	96		202
12	1	349						350
13	12	83						95
14	1	243			16	65		325
15	2	174			4			180
16		198			9	20		227
17	5	146			—5	11		157
18	20	70			13	8		111
19		174			3	8		185
20	93	109		3	17	17		239
21	20	808			31	4		863
22		153	4	23	4	1		185
23	18	13		8	30	335		404
24	8	31			15	848		902
25	9	98			1	5		113
26	21				26	52		99
27			1	29	—5	49		74
28	1	149	33	180	63	65		491
29	1	555	8	335	85	113		1,097
30	2	36			1	2		41
31	12	19		6	6	37		80
32		200	8	171	33	21		433
33	17	27			11	21		76
34		275			33	8		316
35	4	172		1	24	196		397
36		78	10	20	6	5		119
37			849	2,926				3,775
38	478	2,339	16	139		407		3,379
39		195	21					216
40		299		24		3		326
41		2,102						2,102
42		1,617	2,844			1		4,462
43		—302			82	608	—110	278
44	318	1,371		857	56	70		2,672
45	3	147		84				234
46	8	1,172		77				1,257
Total	1,085	15,603	3,794	1,891	955	5,216	110	31,434

Note: Row 43 is unallocated. Row 44 is imports of goods and services. Row 45 is indirect taxes on imported goods and services. Row 46 is indirect taxes on domestic goods and services. This table is comparable to column 43 and columns 45-50 of Table 1 of [6].

TABLE 5

Final Demand by Industry, 1956, Models I, II, and III

(millions of 1949 dollars)

Industry number	1 Final demand Model I	2 Final demand for competitive imports	3 Final demand Model II (1+2)	4 Total competitive imports	5 Final demand Model III (3—4)
1	1,971	79	2,050	157	1,893
2	82		82	10	72
3	70	1	71	4	67
4	877		877	31	846
5	141		141	375	—234
6	86		86	25	61
7	880	13	893	58	835
8	529	3	532	4	528
9	127	9	136	12	124
10	209	19	228	44	184
11	202		202	2	200
12	350	5	355	5	350
13	95		95		95
14	325	18	343	19	324
15	180	8	188	14	174
16	227	8	235	16	219
17	157	2	159	2	157
18	111	12	123	33	90
19	185	12	197	20	177
20	239	47	286	307	—21
21	863	34	897	48	849
22	185	8	193	13	180
23	404	6	410	48	362
24	902	5	907	63	844
25	113	30	143	55	88
26	99	1	100	177	—77
27	74	142	216	164	52
28	491	384	875	713	162
29	1,097	182	1,275	563	716
30	41	17	58	38	20
31	80	6	86	57	29
32	433	150	583	244	339
33	76	24	100	104	—4
34	316	73	389	171	218
35	397	15	412	176	236
36	119	61	180	113	67
37	3,775		3,775	4	3,771
38	3,379	97	3,476	339	3,137
39	216		216		216
40	326		326	7	319
41	2,102		2,102		2,102
42	4,462	1	4,463	9	4,454

The appropriate final demand vector was then aplied to the 1949 inverse matrix for each of the three models to produce backcasts of industry output and of imports for 1956 (in 1949 dollars). This was done both for the forty-two-industry aggregation published in [6] and for the sixteen-industry aggregation.

Table 6 presents the backcasts of the intermediate output of each of the forty-two domestic industries derived from each of the three models. Table 7 presents similar estimates for the sixteen-industry aggregation. Although the models actually backcast the total output of each industry, this includes the final demand which was used to initiate the backcast. In fact therefore the input-output technique backcasts or projects only the intermediate output of each industry (that is, total output less final demand directed towards the domestic industry). For each industry, therefore, the amount of final demand for each model which represented demand for the output of the domestic industry was subtracted from the total backcast output of that industry. This estimate was then compared with an estimate of the "actual" intermediate output of that industry in 1949 dollars derived from the Index of Industrial Production [8] and [1], the Index of Farm Production [7], and Construction in Canada [9], using the 1949 figures in [6, Table 1] as the basis of the projection and subtracting the component of final demand actually directed towards the domestic industry.[6] This procedure of comparing estimates of intermediate output[7] makes apparent the magnitude of the variable being estimated and focuses attention on those industries which produce large quantities of intermediate output (such as the primary iron and steel industry) and where output is projected by the input-output technique as opposed to industries which produce mainly final products (such as the agricultural implements industry) and whose output is implicitly predicted as part of the final demand forecast and then fed into the input-output model as a given.[8]

The standard deviation[9] and a coefficient of variation[10] for the backcasts of intermediate output for each model and each aggregation were then computed. See Table 8.

Generally (with the notable exception of agriculture, wood products, and construction), the tendency was for Model III to underestimate industry output. It is possible that one of the causes of this was an overestimation of competitive imports and a resulting underestimation of the final demand for domestic industries. This is most likely for manufacturing industries where the lack of standard classification and lack of detail makes it difficult to compare domestic products and imports (see section V of this paper). This hypothesis is supported by a rank cor-

---

[6] No independent estimates of output of the services industries were available. The estimates in Hood and Scott [29] stop at 1955 and, moreover, revised data have become available since they made their estimates. Tests of the reliability of the backcasts for the service industries must wait until D.B.S. publishes its extended index of real output.

[7] It should be noted that Models II and III do not require knowledge of the division of final demand into demand for domestically-produced products and for imports. This information is required for Model I so for testing the efficiency of the three models in predicting domestic output, the errors of "prediction" should be judged against the magnitude of the "actual" intermediate output.

[8] In so far, however, that the estimate of final demand directed towards the output of each domestic industry was incorrect, the "actual" intermediate output estimate is wrong by this amount.

[9] The square root of the sum of the squared deviations of the input-output backcast estimates from the "actual" output figures divided by the number of industries. See A. W. Marshall's comments on this [15, pp. 227-30]. This is the numerator of Theil's "inequality coefficient" [47, 66. 31-48] and is identical with Leontief's "standard error of prediction" [33, p. 218].

[10] The standard deviation divided by the average intermediate output to be predicted multiplied by one hundred per cent.

TABLE 6

Comparison of Backcasts of Intermediate Output with "Actual" Intermediate Output, 1956
Forty-two Industries, Detailed Final Demand

(millions of 1949 dollars)

	1	2	3	4
	"Actual" total intermediate output	Backcast of intermediate output minus "actual" intermediate output		
Industry number		Model I	Model II	Model III
1	1,358	213	245	164
2	567	49	52	22
3	62	6	10	9
4	84	44	48	19
5	447	—234	—272	—164
6	114	—39	—39	—57
7	142	—6	—5	—32
8	44	—9	—9	—10
9	12	—3	3	—5
10	15	—3	1	—24
11	300	61	64	56
12	6	0	3	0
13	10	—3	—2	—1
14	3	8	8	11
15	58	11	19	9
16	64	3	4	—2
17	95	—45	—44	—45
18	163	—51	—39	—65
19	81	—21	—14	—25
20	509	50	70	—16
21	26	15	30	4
22	47	10	18	6
23	464	177	184	148
24	601	15	23	—30
25	403	20	26	3
26	409	—19	6	—101
27	2	5	48	4
28	987	—88	30	—289
29	695	—151	—53	—368
30	3	5	9	—16
31	173	84	92	43
32	493	—134	—66	—202
33	397	103	102	117
34	886	—228	—234	—263
35	634	—163	—164	—238
36	135	—24	—6	—54
37	848	356	364	337
38	*			
39	*			
40	635	—119	—115	—139
41	*			
42	*			

* Not available.

TABLE 7

Comparison of Backcasts of Intermediate Output with "Actual" Intermediate Output, 1956
Sixteen-Industry Groups, Detailed Final Demand

(millions of 1949 dollars)

	1	2	3	4
	"Actual" total intermediate output	Backcast of intermediate output minus "actual" intermediate output		
Industry number		Model I	Model II	Model III
1	1,358	196	238	151
2	629	64	69	39
3	198	7	12	—34
4	447	—246	—289	—220
5	749	19	59	—38
6	616	63	113	—25
7	511	191	204	157
8	601	16	24	—32
9	1,569	11	137	—305
10	1,190	—286	—186	—577
11	1,683	—432	—423	—550
12	938	—96	—72	—174
13	848	352	359	334
14	*			
15	635	—118	—115	—138
16	*			

* Not available.

TABLE 8

Standard Deviations of Output Backcasts, 1956

	Forty-two Industries	Sixteen Industries
	standard deviation (millions of 1949 dollars)	
Model I	107	200
Model II	108	204
Model III	130	266
	coefficient of variation (per cent)	
Model I	34.0	23.4
Model II	34.4	23.9
Model III	41.2	31.1

relation between the size of the estimate of competitive imports facing a manufacturing industry (column 4 of Table 5) and the extent to which the Model III backcast underestimated industry output (column 4 of Table 6) which gave a coefficient significantly different from zero at the 1 per cent level. A correction for this would not appear to remove all of the observed backcast error in these industries and so part of the error must be attributed to changing input coefficients.

An underestimation of an industry's intermediate output implies that more of it is being used to produce some other products than in the base period and an overestimation implies the converse.

Models I and II, like Model III, generally underestimated the intermediate output of domestic industries. Although the amount of error is not as great as when Model III is used, this may not be so if the estimates of competitive imports used in formulating final demand for Model III are adjusted downward. The performance of Models I and II in the 1949-56 backcast was very similar. This raises the question whether they were in fact the same model, that is, that the estimates of the coefficients were such that

$$\frac{m_{i1}}{a_{i1}} = \frac{m_{i2}}{a_{i2}} = \ldots = \frac{m_{i42}}{a_{i42}} = \frac{X_i^M}{X_i^D} = m_i$$

for each i. (See [36, pp. 147 ff.].) A comparison of the Model I estimates of X_{ij}^M and Y_i^M with an import input matrix constructed on the assumption that $X_{ij}^M = m_i X_{ij}^D$ and with the final demand for competitive import estimated on a similar assumption revealed some similarities. In some industries (for example, products of petroleum and coal), the similarity of the estimates of a few large coefficients explained the similarity of the backcasts. Many differences were also apparent, however, and, on balance, Model I was significantly different from Model II.

Table 9 presents backcasts of intermediate competitive imports derived from Table 6 for the forty-two-industry Models I and II. Table 10 presents similar estimates for the sixteen-industry Model I. (It will be recalled that in Model III all competitive imports are predetermined.) Estimates of the "actual" competitive imports were derived from [10 (1956)], converted to 1949 dollars, and classified into imports which are for use by final demand sectors without further processing and into intermediate imports. In Model I, the former are predetermined and are excluded from the final demand for the products of domestic industries. Model I therefore only backcasts intermediate imports whereas Model II backcasts total competitive imports.[11]

Since a matrix of competitive imports used by industries in Model I was constructed for 1949 so that each import group was classified both by the industry using the commodity and by the domestic industry producing the similar good, the backcast from Model I could be aggregated to show either the total intermediate imports used by each industry or the total intermediate imports classified by the domestic industry producing the similar good. Tables 9 and 10 are on the latter basis. Model II gives rise only to the latter types of projection unless the assumption is made that all users use imports and domestic goods of a similar type in the same proportions.

Tables 9 and 10 also show a backcast of imports made by using the import coefficients of Model I and Model II applied to the "actual" industry outputs in 1956 rather than to the input-output backcasts of those outputs. It can be seen that in most cases the difference between the two backcasts is negligible. The errors in import backcasts appear to arise, therefore, mainly from changes in im-

_____

[11] Since competitive imports do enter into the final demand of Model II, even though their separate specification is not necessary, it seems appropriate to judge the magnitude of the errors of "prediction" against the level of intermediate competitive imports.

TABLE 9

Comparison of Backcasts of Intermediate Competitive Imports with "Actual" Intermediate
Competitive Imports, Models I and II, 1956 Forty-two Industries, Detailed Final Demand
(millions of 1949 dollars)

	1	2	3	4	5
		Model I		Model II	
Industry number	"Actual" intermediate competitive imports	Input-output backcast minus "actual" imports	"Actual" output backcast minus "actual" imports	Input-output backcast minus "actual" imports	"Actual" output backcast minus "actual" imports
1	78	—19	—19	—45	—51
2	10	—9	—9	—9	—9
3	3	4	4	3	2
4	31	—5	—4	—4	—4
5	375	92	180	130	571
6	25	—15	—14	—14	—15
7	45	—21	—20	—21	—21
8	1	—1	—1	—1	—1
9	3	—2	—2	—7	—8
10	25	—20	—20	—25	—29
11	2	—1	—1	—1	—1
12	0	0	0	—3	—3
13	0	0	0	0	0
14	1	3	3	4	4
15	6	—1	—1	—7	—8
16	8	—4	—4	—10	—10
17	0	0	0	—1	—1
18	21	—7	—6	—15	—11
19	8	—3	—3	—8	—8
20	260	—35	—42	—38	—62
21	14	—10	—11	—25	—27
22	5	—4	—4	—10	—10
23	42	—19	—21	—21	—25
24	58	—28	—27	—29	—30
25	25	—13	—13	—16	—12
26	176	—19	—10	—6	—8
27	22	—1	—2	—43	—91
28	329	—153	—149	—236	—243
29	381	—171	—155	—246	—245
30	21	—17	—17	—20	—23
31	51	—24	—24	—24	—33
32	94	—49	—44	—107	—98
33	80	—6	—5	—4	24
34	98	—8	—3	7	50
35	161	—54	—45	—46	—25
36	52	—25	—24	—41	—39
37	4	—3	—3	—3	1
38	242	37	61	4	2
39	0	0	0	0	0
40	7	—5	—5	—5	—5
41	0	0	0	0	0
42	8	—5	—5	—5	—3

TABLE 10

Comparison of Model I Backcasts of Intermediate Competitive Imports
with "Actual" Intermediate Competitive Imports, 1956
Sixteen-Industry Groups, Detailed Final Demand
(millions of 1949 dollars)

Industry number	1 "Actual" intermediate competitive imports	2 Input-output backcast minus "actual" imports	3 "Actual" output backcast minus "actual" imports
1	78	—18	—17
2	13	—5	—6
3	56	—19	—18
4	375	55	124
5	91	—45	—45
6	282	—48	—54
7	47	—22	—24
8	58	—28	—27
9	556	—186	—178
10	497	—237	—215
11	280	—66	—49
12	178	—60	—59
13	4	—3	—3
14	242	32	50
15	7	—5	—5
16	8	—5	—5

port coefficients rather than from the errors in projecting the outputs of domestic industries. There may have been a tendency to substitute imports for domestic goods since the use of constant ratios led to an underestimate of competitive imports (except for the petroleum, natural gas, and coal group). Unfortunately, 1949 was a bad year to choose for the estimation of market share parameters because of the import restrictions which were in effect that year.(See, [6, p. 19].) Some increase was to be expected simply as a result of the lifting of the controls. The increase in the market share was, however, more than would be expected as a result of this factor.

Another factor that may contribute to an explanation of the backcast errors is that imports may have been used to supplement rather than to replace domestic output at high-level employment and the increase in imports in some years may simply reflect the reaching of capacity output by some domestic producers [30, pp. 40-41] and [43, p. 24]. Moreover, some of the goods classed as competitive may in fact be of a different kind than that produced domestically and should have been classed as non-competitive. (See section V.)

The standard deviations on an absolute and relative basis for the three input-output backcasts of competitive imports shown in Tables 9 and 10 are given in Table 11. The similar measures for the backcasts using "actual" industry output figures are shown in Table 12.

TABLE 11

Standard Deviation of Input-Output Import Backcast, 1956

	Forty-two Industries	Sixteen Industries
	standard deviation (millions of 1949 dollars)	
Model I	42	83
Model II	61	*
	coefficient of variation (per cent)	
Model I	63.6	47.9
Model II	60.5	*

* not computed

TABLE 12

Standard Deviation of "Actual Output" Import Backcast, 1956

	Forty-two Industries	Sixteen Industries
	standard deviation (millions of 1949 dollars)	
Model I	47	82
Model II	107	*
	coefficient of variation (per cent)	
Model I	71.2	47.3
Model II	105.9	*

* not computed.

IV. BACKCASTS FROM VARIOUS FINAL DEMAND VECTORS, 1950–1958

Backcasts using less information on the components of final demand than was used for the 1956 backcasts of the previous section were made for the year 1956 to see what improvement in the estimates was obtained from using increasing amounts of information on the components of final demand and by using the input-output technique instead of a simpler projecting technique. These various methods (except for the detailed final demand vector which was only constructed for the year 1956) were also used for the years 1950-55 and 1957-58 in order to form some impressions as to how the errors of prediction changed over time and to suggest the possibility of incorporating adjustments in the projection procedure based on observed errors of estimate of previous years. Such a procedure might lengthen the useful life of an input-output matrix. A further reason for using less detailed final demand information was to see whether forecasts of the variables published annually by D.B.S. in [3, Table 5] were sufficient to make reliable projections using the input-output technique. In addition to serving as a method of evaluating the input-output backcasts, it was also hoped that the study of the period 1949-58 would throw some light on the extent to which changes in the level of imports over the period could be attributed to changes in the composition of final demand.

The input-output technique is designed to take into account changes in the composition of final demand. A proportionate change in all components will change all industry outputs in the same proportions from the base year. The technique must therefore give a better estimate of industry outputs than that resulting from the assumption that all outputs have changed in the same proportion if it is to be useful when a given change in final demand occurs. The various assumptions about the available data on final demand in the years 1950-58 were as follows:

Alpha: Only the total gross domestic product at market prices in 1949 dollars is known. (See line 15 of Table 13.) All industry outputs are therefore changed in the same proportion and imports are assumed to be a fixed proportion of G.D.P. This does not involve use of the inverse matrix and is therefore a control for the efficiency of input-output projections.

Beta: The totals of the following five components of G.D.P. are known in 1949 dollars. (See lines 1, 5, 9 and 13 of Table 13. Inventory change is shown in [3 and 4, Table 5].)

   (1) personal expenditure on consumer goods and services,
   (2) government expenditure on goods and services,
   (3) business gross fixed capital formation,
   (4) value of the physical change in inventories,
   (5) exports of goods and services.

The industrial distribution of each of the various categories of domestic demand and the import content are assumed to be the same as in 1949, except for inventory change. Inventory change was computed on the basis of the average of the percentage distributions of the absolute value of the change in 1949 and 1956.

Gamma: The totals of the following twelve components of G.D.P. in 1949 dollars are known.[12] (See lines 2, 3, 4, 6, 7, 8, 10, 11, 12 and 13 of Table 13. The two components of inventory change are given in [3 and 4, Table 5].)

   (1) personal expenditure on non-durable consumer goods
   (2)   ,,   ,,   ,, durable   ,,   ,,
   (3)   ,,   ,,   ,, services
   (4) government expenditure on current goods and services
   (5)   ,,   ,,   ,, machinery and equipment
   (6)   ,,   ,,   ,, new construction
   (7) business   ,,   ,, imported machinery and equipment
   (8)   ,,   ,,   ,, domestic machinery and equipment
   (9)   ,,   ,,   ,, new construction
  (10) value of the change in non-farm business inventories
  (11) farm inventories and grain in commercial channels
  (12) exports of goods and services

Because of the volatility of imported machinery and equipment, component (7) was estimated directly from [10] and converted to 1949 dollars. The industrial

---

[12] In a paper "L'Ajustement périodique des systemes de relations inter-industrielles, Canada, 1949-1958" by the present authors in *Econòmetrica* [50], this type of final demand estimate is referred to as "deuxième approximation, base 1949."

TABLE 13

Changes in Components of Final Demand, 1949 Dollars, 1949-1958

Final Demand Component	1949	1950	1951	1952	1953	1954	1955	1956	1957	1958
	millions of dollars	Index 1949 = 100								
1. Personal Expenditure on Consumer Goods and Services	10,923	106.6	108.2	115.7	122.1	125.0	134.2	142.8	147.0	151.7
2. Non-durable goods	6,288	104.5	106.0	111.6	116.9	119.8	126.6	134.4	139.4	142.8
3. Durable goods	1,146	125.0	113.2	133.2	151.6	150.4	180.3	192.8	185.5	188.2
4. Services	3,489	104.2	110.4	117.3	121.9	125.9	132.8	141.6	148.0	155.7
5. Government Expenditure on Goods and Services	2,127	105.4	131.9	165.3	165.4	160.6	167.5	178.4	179.5	187.3
6. Current goods and services	1,620	103.7	132.1	164.8	169.6	165.2	170.8	177.1	175.9	186.9
7. Machinery and equipment	55	105.5	112.7	158.2	160.0	149.1	130.9	138.2	140.0	165.5
8. Construction	452	111.5	133.6	167.9	150.7	145.4	160.2	187.8	196.9	191.4
9. Business Gross Fixed Capital Formation	3,032	104.5	108.9	118.3	129.5	122.8	130.7	161.3	167.7	156.1
10. Imported machinery and equipment	575	94.1	116.0	124.0	136.9	124.3	137.2	171.1	160.2	132.9
11. Domestic machinery and equipment	743	108.3	112.1	121.5	124.9	106.3	103.4	132.0	142.5	117.8
12. Construction	1,714	106.2	105.1	115.1	129.0	129.4	140.3	170.7	181.1	180.6
13. Exports of Goods and Services	3,938	99.4	108.7	119.4	118.3	113.8	122.5	132.0	132.6	134.6
14. Imports of Goods and Services	3,463	108.3	124.1	130.7	141.8	133.8	153.8	179.6	176.4	165.1
15. Gross Domestic Product at Market Prices	16,650	107.1	113.1	121.6	126.1	122.7	133.3	144.9	145.3	147.1

Source: [3 and 4 (1960), Table 5] with adjustments to bring figures to a domestic output basis. In all years, personal expenditure on gas and electricity is included in personal expenditure on services. The estimate of business expenditure on imported machinery and equipment was calculated from [10] and converted to 1949 dollars using import price indexes.

TABLE 14

Comparison of Various Backcasts of Total Industry Output, 1956
(millions of 1949 dollars)

	1	2	3	4	5
			Error (Backcast minus "Actual")		
			Model I	Model I	Model I
Industry number	"Actual" output	Alpha backcast	Beta backcast	Gamma backcast	Delta backcast
1	3,329	176	448	414	213
2	649	—19	52	48	49
3	132	40	28	23	6
4	961	—21	—78	—75	44
5	588	—331	—302	—309	—234
6	200	—57	—59	—59	—39
7	1,022	8	54	—5	—6
8	573	68	59	26	—9
9	139	43	24	16	—3
10	224	—10	83	—34	—3
11	502	124	159	141	61
12	356	28	20	—2	0
13	105	20	13	7	—3
14	328	5	1	—16	8
15	238	76	73	55	11
16	291	21	30	13	3
17	252	0	—9	—23	—45
18	274	—14	—1	—9	—51
19	266	70	66	59	—21
20	748	183	138	139	50
21	889	214	191	129	15
22	232	9	13	53	10
23	868	51	78	98	177
24	1,503	88	25	19	15
25	516	33	56	46	20
26	508	—40	3	2	—19
27	76	179	174	152	5
28	1,478	—115	21	—33	—88
29	1,792	—136	—115	—84	—151
30	44	35	30	49	5
31	253	61	71	83	84
32	926	—220	—174	—145	—134
33	473	—113	—104	—95	—103
34	1,202	—391	—323	—339	—228
35	1,031	—170	—166	—183	—163
36	254	—2	6	3	—24
37	4,623	—277	155	356	356
38	*				
39	*				
40	961	—255	—244	—321	—119
41	*				
42	*				

* Not available.

distribution and import content of each of the categories of domestic demand (other than inventory change) are assumed to be the same as in 1949. The change in non-farm inventories was calculated using the average of the percentage distributions of the absolute value of non-farm inventory change in 1949 and in 1956.

Delta: The industrial distribution of each of the eleven domestic components is estimated using supplementary data for the year for which the backcast was made. This was done only for 1956 and these backcasts were discussed in the previous section. (See Tables 6, 7, 9, 10, and 14.)

Epsilon: The actual output of each industry is known. Imports are then estimated on the assumption that the 1949 ratio of imports to domestic output is constant. This procedure removes the intermediate step of estimating industry outputs and enables a direct check on the stability of the import coefficients. (See Tables 9 and 10.)

Table 14 compares various forty-two industry Model I backcasts (designated by the Greek letters above) of industry output in 1956 indicating the way in which errors behave as the amount of information on final demand increases. Table 15 gives the various standard deviations. The lower standard deviation for

TABLE 15

Standard Deviations of Various Output Backcasts, 1956

	Standard deviation (millions of 1949 dollars)
Alpha	138
Model I Beta	138
Model I Gamma	146
Model I Delta	107

the Delta backcast suggests that the input-output technique only becomes superior to alternative simpler methods when final demand is specified in detail.

Table 16 gives an indication of the behaviour of the errors of a specific type of backcast (the Model I Gamma backcast) over time. This backcast incorporates the maximum amount of information on final demand which was utilized for all nine years. It might be noted that these tables refer to *total* industry output, rather than intermediate output, since the final demand for each industry is not a given but is based on the 1949 ratio of the industry's final output of the relevant category of final product to the total final demand in that expenditure category.[13]

Various Model I backcasts of total imports from 1950 to 1958, are shown in Table 17. These backcasts are all based on the forty-two-industry aggregation. The superior performance of the Delta backcast for 1956 (the only year for which it was made) is explained by the fact that this is the only backcast which utilized current data to estimate all components of predetermined competitive imports. The standard deviations for the backcasts are as shown in Table 18.

---

[13] The coefficient of variation in section III of this paper have average *intermediate* output (or imports) in their denominator and are not comparable with coefficients of variation which could be calculated from data in this section which would have average *total* output in the denominator.

TABLE 16

Model I Gamma Backcasts of Total Industry Output as Percentage
of "Actual" Industry Output, 1950-1958
Forty-two Industries, Twelve Components of Final Demand, 1949 Dollars

Industry number	1950	1951	1952	1953	1954	1955	1956	1957	1958
					per cent				
1	103	102	102	105	118	117	112	126	117
2	96	85	102	107	97	97	107	119	125
3	88	96	108	107	91	106	117	113	107
4	97	100	106	100	88	86	92	84	79
5	104	87	78	71	61	53	47	45	47
6	78	73	75	78	66	66	70	68	65
7	106	111	97	104	98	96	100	101	96
8	109	104	110	107	100	100	104	102	99
9	81	89	100	99	84	102	112	105	96
10	89	79	74	84	66	67	85	76	72
11	113	111	119	120	110	128	128	122	108
12	101	97	98	96	98	100	99	107	109
13	103	111	108	106	112	103	107	103	103
14	102	99	93	91	92	94	95	94	93
15	96	123	113	121	118	123	123	125	122
16	104	101	101	101	98	100	104	101	97
17	99	110	100	95	91	89	91	83	77
18	97	95	101	98	101	94	97	98	102
19	112	120	114	114	123	123	122	124	128
20	96	97	111	112	125	112	118	117	125
21	103	105	98	101	107	110	114	117	120
22	110	89	114	112	108	118	123	120	117
23	98	97	106	102	100	98	111	126	120
24	94	102	106	104	95	97	101	103	104
25	104	110	119	116	109	110	109	110	116
26	104	93	97	110	128	100	100	104	118
27	122	126	120	164	208	219	300	304	263
28	105	96	102	108	101	99	98	101	107
29	100	85	78	76	84	88	95	97	103
30	111	115	156	148	164	194	211	207	200
31	108	94	105	114	121	122	133	146	144
32	101	95	101	90	84	79	84	88	87
33	94	92	96	90	82	77	80	80	73
34	101	94	90	86	75	73	72	67	68
35	97	92	96	87	77	78	82	76	74
36	103	94	101	92	94	99	101	98	93
37	102	103	106	106	105	106	108	107	108
38	*								
39	*								
40	83	79	78	78	73	68	66	62	58
41	*								
42	*								

* Not available.

TABLE 17

Comparison of Various Model I Backcasts of Total Imports,* 1950-1958
(millions of 1949 dollars)

	1	2	3	4	5
			Error (backcast minus actual)		
			Model I	Model I	Model I
	Actual	Alpha	Beta	Gamma	Delta
Year	imports	backcast	backcast	backcast	backcast
1950	3,752	—50	—82	—80	
1951	4,299	—347	—467	—402	
1952	4,525	—264	—439	—490	
1953	4,909	—470	—618	—559	
1954	4,632	—353	—449	—447	
1955	5,325	—648	—798	—779	
1956	6,220	—1,136	—1,273	—1,102	—742
1957	6,108	—975	—1,062	—1,063	
1958	5,717	—633	—746	—863	

* Domestic basis, i.e., excluding interest and dividends paid abroad.

Note: Model I Delta Backcast available only for 1956. All backcasts assume the "unallocated" component of final demand to be the same as in 1949.

TABLE 18

Standard Deviations of Various Import Backcasts, 1950-58

	Standard deviation (millions of 1949 dollars)
Alpha	632
Model I Beta	741
Model I Gamma	716

V. CONCLUSION AND COMMENTS ON INTER-INDUSTRY ANALYSIS

This paper set out (*a*) to compare three alternative treatments of competitive imports in input-output models and (*b*) to compare one input-output model (Model I) utilizing a detailed final demand vector for the backcast year with the same model using less final demand detail and with a simple non-input-output model (Alpha model). Since projections for future years are difficult to evaluate because they depend upon a correct forecast of final demand, backcasts for past years for which final demand was known were used to attempt to isolate the errors attached to input-output projections. The tests among the three models were not conclusive, however, although sufficient evidence was established that both import coefficients and technical input coefficients changed substantially over the period 1949-56. The tests were hampered by statistical difficulties which will be discussed below.

When the input-output technique was compared with a simpler technique, it was concluded that, providing a detailed final demand vector was given, the input-output technique was slightly better. Input-output analysis is designed, however, to project the effects on intermediate output of large shifts in the composition of final demand. The shifts that took place in Canada from 1949 through 1958 were not great, as can be seen from Table 13. All components increased at rates varying from about 20 to 90 per cent. A situation where some components were decreasing might have distinguished more clearly between the power of input-output projections and simpler methods such as the Alpha projection.

It should also be emphasized that the projection techniques used in this paper and by most other input-output researchers were quite mechanical. All coefficients were treated as constants even though some information was readily accessible which could have been used to update the coefficients to take into account known changes in technology or sources of supply. It is relatively simple for the analyst to change a few "key" coefficients in a model in order to improve the results, provided he has high-speed computing equipment available to him. The large amount of detail in an input-output table makes it easy to insert the effect of known changes in an industry into the model, whereas in an aggregative model the aggregative effect of such a change would likely be either ignored or guessed. As Chenery and Clark have said, "The real limitations (of the inter-industry approach) are the analyst's understanding of what deviations from the Leontief proportionality assumption are appropriate and his willingness to take pains to incorporate them quantitatively into the model" [18, p. 266]. Moreover, "the use of an inter-industry model does not preclude separate analysis of important sectors, but rather serves to make the sector results consistent with each other" [18, p. 335]. Indeed, this was the purpose of the Caves and Holton use of an inter-industry projection of industry outputs for 1970 [17].

With respect to the treatment of the relations between imports and domestic output, it might be noted that the construction of a separate import matrix as in Model I has also been followed for Italy with some success (cited in [18, p. 264, n. 2]) and for analyses of Columbia and Argentina (cited in [18, p. 195]). Most researchers, have however, favoured either a Model II approach, utilizing a variable import coefficient for each industry which faces competitive imports, or Model III which treats imports as predetermined. Either of these two approaches can allow for time-trends in substitution between sources of supply, expected changes in comparative costs, and limitations of domestic capacity or foreign supply [18, pp. 154-5] and [13, p. 48].

This raises the appropriateness of the assumption of constant-dollar market shares used in this paper, the m_i. While there may appear to be some justification for ignoring the effect of changes in relative prices upon technical input coefficients as long as the price changes are not great and the time period is short, there is no such justification when it affects the choice between two suppliers of a similar commodity. As a first approximation to a plausible theory of market shares, a constant current-dollar ratio might be suggested. When using such a ratio, changes in relative prices would produce offsetting changes in quantities. A look at the ratios of competitive imports to domestic output (m_i) for some of the major import-competing industries dispels, however, any notion of constancy in the ratios over time. (See Table 19.)

A more reasonable approach might be to use flexible import ratios which incorporate the latest information on quotas, tariffs, relative prices, domestic produc-

TABLE 19

Ratio of Competitive Imports to Domestic Output, 1949 and 1956

Industry	1949 ratio	1956 ratio in 1949 dollars	1956 ratio in 1956 dollars
1	.032	.049	.051
5	1.606	.641	.607
20	.409	.512	.343
26	.367	.383	.351
27	.977	2.158	1.831
28	.333	.507	.443
29	.197	.339	.264
30	.385	.974	.780
31	.105	.248	.156
32	.169	.278	.262
33	.281	.229	.205
34	.193	.149	.124
35	.163	.186	.174
36	.297	.454	.331

tion capacities, etc. Using Model II one would use a first-approximation import ratio different from the base-year m_i to divide the estimated final demand for a commodity between imports and domestic production, feed the information into the model and using the same ratio (or possibly a different one to allow for difference in the price-elasticity of demand for final goods and intermediate goods produced within the same industry group), and derive an estimate of total imports and domestic output. The latter estimate can then be compared with productive capacities and other factors and, if necessary, a new import ratio calculated and the revised solution produced. With electronic computing equipment available, a series of such iterations is feasible.

Indeed, under some situations a combination of the three models may be used in order to take account of various pieces of information about the year for which the projection is being made. It would appear, therefore, that when flow tables are being constructed, if the data are available, the table corresponding to Model I should be constructed. This requires a matrix of inter-industry flows for domestically produced goods and a separate matrix for competitive imports. With these data available the inverse matrices for Models I, II and III can be constructed and the analyst is then free to select whichever Model suits his data and objectives.

Data problems still, however, hamper the inter-industry analyst. The D.B.S. 1949 inter-industry flow table has varying degrees of reliability associated with various cells in the table [6, Table 12] and one data gap affected nearly all estimates of commodity flows. This is the lack of information upon transportation, storage, trade margins, and taxes entering into the difference between the value of the good as it leaves the producing establishment. The adoption of a system of either producers' prices or purchasers' prices does not solve this problem; it merely transfers the errors to different cells of the table. A second major problem associated with the construction of the basic table is the estimation of goods and non-factor services, other than materials, fuel and electricity, used by manufacturing industries. If such data could be estimated on an establishment basis, the gross domestic product originating in an industry could be estimated by subtracting intermediate input from total output. This would provide a check on estimates made by the income-originating approach as well as on estimates of the intermediate output of service industries and of industries producing supplies for use other than in manufacturing operations. The over-all reliability of the table would thus be improved.

In making the backcasts from the models of this paper, especially Models II and III, the estimation of competitive imports was crucial. The comparison of the backcasts of competitive imports from Modell II with the "actual" assumes the latter correct. This involves correctly identifying the competitive imports from data in [10], correctly relating these to the proper domestic industry, and deflating by an appropriate price index to convert to 1949 dollars. Some of the difficulties resulting from the lack of detailed information on imports may be illustrated by the chemicals group. Two commodity subgroups which are difficult to compare with domestic production are "plastics" and "drugs, medicinal and pharmaceutical preparations." The proportion of these two commodities to total imports of commodities similar to those produced in Canada by the chemicals and allied products industry has increased in current dollars from 40 per cent in 1949 to 53 per cent in 1958. Virtually all imports of these goods were classed as competitive imports for this study. Moreover, the "n.o.p." item for chemicals was prorated into competitive and non-competitive imports on the assumption that two-thirds of this item were in the latter category. In 1958 this accounted for 48 per cent of the estimated non-competitive imports of chemicals and allied products. Until the D.B.S. Standard Commodity Classification is implemented uniformly for both imports and domestic production, there seems little hope of improving these estimates.

REFERENCES AND SELECT BIBLIOGRAPHY

A. *Publications of the Dominion Bureau of Statistics* (Ottawa: Queen's Printer)

1. No. 11-003, *Canadian Statistical Review* (monthly)
2. No. 13-001, *National Accounts, Income and Expenditure* (quarterly)
3. No. 13-201, *National Accounts, Income and Expenditure* (annual)
4. No. 13-502, *National Accounts, Income and Expenditure, 1926-1956* (1958)
5. No. 13-510, *Inter-Industry Flow of Goods and Services, Canada 1949* (1956)
6. No. 13-513, *Supplement to the Inter-Industry Flow of Goods and Services, Canada, 1949* (1960)
7. No. 21-203, *Index of Farm Production* (annual)
8. No. 61-502, *Revised Index of Industrial Production, 1935-1957* (1959)
9. No. 64-201, *Construction in Canada* (annual)
10. No. 65-203, *Trade of Canada*, vol. III, *Imports* (annual)

B. *Other Publications*

11. Adams, A. A., and I. G. Stewart, "Input-Output Analysis: An Application," *Economic Journal*, vol. LXVI, Sept., 1956, pp. 442-54.
12. Arrow, K., and M. Hoffenberg, *A Time Series Analysis of Interindustry Demands* (Amsterdam, 1959).

13. Aukrust, O., and Secretariat of UN Economic Commission for Europe, "Input-Output Tables: Recent Experience in Western Europe," *Economic Bulletin for Europe*, vol. VIII, May, 1956, pp. 36-53.

14. Balderston, J. B., and T. M. Whitin, "Aggregation in the Input-Output Model," in O. Morgenstein, ed., *Economic Activity Analysis* (New York, 1954), pp. 79-126.

15. Barnett, H. J., "Specific Industry Output Projections," National Bureau of Economic Research Studies in Income and Wealth, vol. XVI, *Long-Range Economic Projections* (Princeton, 1954), pp. 191-226, and "Comments" by A. W. Marshall and S. Lebergott, *ibid.*, pp. 227-232.

16. Caves, R. E. "The Inter-Industry Structure of the Canadian Economy," *Canadian Journal of Economics and Political Science*, vol. XXII, Aug., 1957, pp. 313-30.

17. Caves, R. E., and R. H. Holton, *The Canadian Economy: Prospect and Retrospect* (Cambridge, Mass., 1959).

18. Chenery H. B., and P. G. Clark, *Interindustry Economics* (New York, 1959).

19. Christ, C. F. "A Review of Input-Output Analysis," National Bureau of Economic Research Studies in Income and Wealth, vol. XVIII, *Input-Output Analysis: An Appraisal* (Princeton, 1955), pp. 137-169, and "Comments" by M. Friedman and P. N. Ritz, *ibid.*, pp. 169-82.

20. Cornfield, J., W. D. Evans, and M. Hoffenberg, "Full Employment Patterns, 1950," *Monthly Labour Review*, vol. LXIV, 1947, pp. 163-90 and 420-32.

21. Dorfman, R., P. A. Samuelson, and R. M. Solow, *Linear Programming and Economic Analysis* (New York, 1958).

22. Evans, W. D., "The Effect of Structural Matrix Errors on Inter-Industry Relations Estimates," *Econometrica*, vol. XXII, Oct., 1954, pp. 461-80.

23. Evans, W. D., "Input-Output Computations," in T. Barna, ed., *The Structural Interdepence of the Economy* (New York, 1956).

24. Evans, W. O., and M. Hoffenberg, "The Interindustry Relations Study for 1947," *Review of Economics and Statistics*, vol. XXXIV, May, 1952, pp. 97-142.

25. Georgescu-Roegen, N., "Leontief's System in the Light of Recent Results," *Review of Economics and Statistics*, vol. XXXII, Aug., 1950, pp. 214-22.

26. Giersch, Herbert, "The Acceleration Principle and the Propensity to Import," *International Economic Papers*, no. 4, 1954, pp. 197-229.

27. Heady, E. O. and W. Candler, *Linear Programming Methods* (Ames, iowa, 1958), chaps. 11 and 14.

28. Hoffenberg, M., "Prices, Productivity, and Factor Return Assumptions in Long-Range Economic Projections" in American Statistical Association, *Proceedings of the Business and Economic Statistics Section, 1955-56*, pp. 16-19.

29. Hood, Wm. C., and A. D. Scott, *Output, Labour and Capital in the Canadian Economy* (Ottawa: Queen's Printer, for the Royal Commission on Canada's Economic Prospects, 1958).

30. Humphrey, D. D., *American Imports* (New York, 1955).

31. Johansen, Leif, *A Multi-Sectoral Study of Economic Growth* (Amsterdam, 1960).

32. Klein, L. R., "On the Interpretation of Professor Leontief's System," *Review of Economic Studies*, vol. XX, 1952-53, pp. 131-6.

33. Leontief, W. W., *The Structure of American Economy, 1919-1939* (2nd ed., New York, 1951).

34. Leontief, W. W., "Structural Change", in Leontief, ed., *Studies in the Structure of the American Economy* (New York, 1953), 17-52.

35. Malinvaud, E., "Aggregation Problems in Input-Output Models," in T. Barna, ed., *The Structural Interdependence of the Economy* (New York, 1956), pp. 187-202.

36. Modlin, C. P., and G. Rosenbluth, "The Treatment of Foreign and Domestic Trade and Transportation Charges in the Leontief Input-Output Table," in O, Morgenstein, ed., *Economic Activity Analysis* (New York, 1954), pp. 129-99.

37. Matuszewski, T. I., "On Some Input-Output Computations," *Journal of the Royal Statistical Society*, vol. A CXXIII, 1960, pp. 195-9.

38. Rasmussen, P. N., *Studies in Inter-Sectoral Relations* (Amsterdam, 1953).

39. Rhomberg, Rudolf R., "Canada's Foreign Exchange Market," *International Monetary Fund Staff Papers*, vol. VII, April, 1960, pp. 439-56.

40. Sawyer, J. A., "The Measurement of Inter-Industry Relationship in Canada," *Canadian Journal of Economics and Political Science,* vol. XXI, Nov., 1955, pp. 480-97.
41. Sevaldson, P., "Norway," in T. Barna, ed., *The Structural Interdependence of the Economy* (New York, 1956).
42. Slater, D. W., "Changes in the Structure of Canada's International Trade," *Canadian Journal of Economics and Political Science,* vol. XVI, Feb., 1955, pp. 1-19.
43. Slater, D. W., *Canada's Imports* (Ottawa: Queen's Printer, for the Royal Commission on Canada's Economic Prospects, 1957).
44. Steinthorson, D. H., "Notes on Inter-Industry Studies," *Canadian Journal of Economics and Political Science,* vol. XXI, Nov., 1955, 533-34.
45. Stone, Richard, *Input-Output and National Accounts* (Paris: Organization for European Economic Co-operation, 1961).
46. Stone, Richard, and Giovanna Croft-Murray, *Social Accounting and Economic Models* (London, 1959).
47. Theil, H., *Economic Forecasts and Policy* (Amsterdam, 1958).
48. United States of America, Mutual Security Agency, Special Mission to Italy for Economic Cooperation, *The Structure and Growth of the Italian Economy* (Rome, 1953).
49. Wonnacott, R. J., *Canadian-American Dependence: An Interindustry Analysis of Production and Prices* (Amsterdam, 1961).
50. Matuszewski, T. I., P. R. Pitts, and J. A. Sawyer, "L'Ajustement periodique des systèmes de relations inter-industrielles, Canada, 1949-1958," *Econometrica,* vol. XXXI, Jan.-April, 1963, pp. 90-110.
51. Matuszewski, T. I., P. R. Pitts, and J. A. Sawyer, "Alternative Treatments of Imports in Input-Output Models: A Canadian Study," *Journal of the Royal Statistical Society,* vol. A CXXVI, no. 3, 1963.

Discussion

S. J. MAY

This paper consists, among other things, of an evaluation of the efficacy of the input-output methodology for purposes of forecasting the industrial distribution of total output, and of competitive imports in particular.

The aims of the paper, so far as I can gather from the text, were as follows: (i) To find out whether, and for how long, input-output relations remain sufficiently stable to make reliable predictions possible. (ii) To test the hypothesis that technical coefficients remain constant while instability results from shifts in sources of supply—that is, that they are the result of market or behaviour (as distinct from technological) phenomena. (iii) To compare the predictive capability of the input-output scheme with that of other simpler techniques.

With regard to stability of the input-output structure the authors establish, I believe, that it is not very stable. They establish this not by following changes in the structure, but by a consideration of the results of instability, namely, large errors in the results of prediction exercises.

With regard to testing the hypothesis that instability results from shifts in sources of supply, the authors arrive at the conclusion that instability does derive from changes in market behaviour (by considering the predictions of competitive imports when actual or observed industry outputs were used). However, I think they could conclude from their tests that the technological structure as measured by input-output is also unstable (Model III results).

With regard to the predictive capability of input-output and that of other simpler techniques, the comparisons are inconclusive.

Each of the models was solved for the year 1956 using varying amounts of required information with respect to the commodity distribution of final demand. The exogenous content of total outputs was deducted from solutions of total outputs and the resulting solutions of intermediate outputs were tabulated, industry by industry, and these latter were used to compute statistics reflecting the goodness of fit. The coefficients of variation for the three models are sizable indeed.

While I was reading the earlier part of their paper, and before I had seen the measures of statistical fit, I already had formed the opinion that the authors selected an extremely harsh test for their models. First, they selected the year 1956 for their tests, a period seven years removed from the 1949 base and a year of much higher activity. Second they considered only intermediate outputs for analysis of goodness of fit.

Now it may be true that the competing forecasting techniques the authors had

C.P.S.A. Conference on Statistics, 1961, *Papers*. Printed in the Netherlands.

in mind do not require a detailed commodity break-down of final demand. However, if the commodity break-down of final demand is made part of the predictive technique of input-output—fixed proportions of a total, for example—the forecast errors associated with the industrial distribution of total output could fairly be compared to the industrial distribution of output arrived at other predicting techniques. I suspect that in this case the input-output technique would not come off quite so badly (see, for example, the authors' gamma forecasts).

Competitive imports predicted by Models I and II are also wide of the mark. However, when actual outputs were used to re-solve for competitive imports the Model I results seem to be significantly better than those of Model II.

The standard error and coefficient of variation are computed for the solution of total intermediate supply of the commodities by Model II. However, no comparable statistics are given for predictions of total intermediate supply by Model I. Considering that the error in predicted total supply of a commodity is the algebraic sum of the error in predicted intermediate domestic supply and the error in predicted competitive imports, and that these errors seem to be inversely correlated with respect to sign in Models I and II, it would not be surprising if the coefficient of variation for the predictions of total intermediate supply by Model I also showed some improvement over the prediction for intermediate domestic output. In Model II, the authors have substituted a different combination of rigid proportionalities than for Model I and the results hardly justify any claim that Model II is superior.[1]

Turning to the particular constructions of Models I and II I sensed that the authors held Model II in rather higher regard than Model I. I am not disposed to the same opinion. In Model I, the various competitive import commodities are related to the output of consuming industries in a highly detailed, rigid structure. This is certainly not in its favour, but at the same time the observed inputs of competitive import commodities in the base period are something more than a statistical accident. Also, there is a certain logic in tying all inputs to the activity level in the consuming industry. Nevertheless, I would agree that the combining of commodity coefficients for both imports and domestic shipments has a greater appeal from the technological point of view and in this respect one would expect more stability in the Model II-type technological coefficients.

However, in Model II the competitive imports are related directly to the competing domestic industry whereas in Model I the level of competitive commodity imports rides up and down in sympathy with the fortunes of the consuming industries. It is true that competitive imports and total domestic output both might be expected to be positively related to final demand for the particular commodity, but any process of substitution would require that the substitution of competitive imports would ride up with the *mis*fortunes of the competing industry. Thus my comment in passing is that the particular choice of proportionalities in Model II hardly captures the essence of behaviouristic substitution of consuming industries, as buyers in a market. As a result, in the case of Model II, it is possible to construct special situations in which the Model II solution could imply some patent absurdities.

[1] Editors' note: The calculations referred to in this paragraph have been deleted from the revised version of the paper published in this volume. Subsequent calculations verified Mr. May's prediction.

The inter-industry flows in Model II combine both domestic and import shipments of a particular commodity and no break-down of the aggregate into domestic and import shipments is ordinarily available in a prediction exercise. However, it is possible to rig a particular situation in which we can sort out the import and domestic content of an inter-industry flow, and in some cases the domestic shipments content of an inter-industry flow can turn out to be negative for a large range of final demands for the particular commodity. This, of course, is substitution with a vengeance. I do not know if this is a serious problem for Model II, but of course we can seldom find out. There are ordinarily too many unknowns compared to the number of equations required to find out.

I should stress that the situation alluded to here was special, or rigged, only in the sense that sufficient restrictions were added so that I could separate the contents of an implied inter-industry flow. The implied negative domestic or import shipments for all I know might be a problem that is rampant in the general Model II situation when there is uneven growth in the commodity distribution of final demand from the base period to the prediction period.

Turning back now to Model I as a technological structure I was very gratified to find the gamma-type solutions covering the years 1950 to 1958, which are contained in Table 10. I would say that the solution vectors for 1950-51-52 provide a less demanding (and perhaps more realistic) test of the efficacy of the technological inter-industry model. After all, if the input-output scheme is to become a serious forecasting tool we could expect that the structure would be updated more frequently. Looking at the 1952 solution the results are hardly outstanding, but they are not altogether bad.

I should like to conclude by thanking the authors for producing this paper on the input-output scheme using the Canadian data. I have only a vague idea of the amount of labour involved in assembling the relevant data for these analyses, but I am sure it must be prodigious, and I want to congratulate the authors for making these results available.